PENGUIN PASSNOTES

Biology

Trevor Day was educated at Nottingham University where he obtained a
B.Sc. in Zoology. He took his M.Sc. in Marine Biology at the University
of North Wales, Bangor. This was followed by a period of research funded
by the United Nations Development Programme in Egypt. He has also
taught biology at a number of schools in this country.

PENGUIN PASSNOTES

Biology

TREVOR DAY
ADVISORY EDITOR: STEPHEN COOTE, M.A., PH.D.

PENGUIN BOOKS

Penguin Books Ltd, Harmondsworth, Middlesex, England
Viking Penguin Inc., 40 West 23rd Street, New York, New York 10010, U.S.A.
Penguin Books Australia Ltd, Ringwood, Victoria, Australia
Penguin Books Canada Ltd, 2801 John Street, Markham, Ontario, Canada L3R 1B4
Penguin Books (N.Z.) Ltd, 182–190 Wairau Road, Auckland 10, New Zealand

First published 1985

Made and printed in Great Britain by
Richard Clay (The Chaucer Press) Ltd, Bungay, Suffolk
Filmset in Monophoto Times Roman by
Northumberland Press Ltd, Gateshead

*The publishers are grateful to the following Examination Boards for
permission to reproduce questions from examination papers used in
individual titles in the Passnotes series:*

*Associated Examining Board, University of Cambridge Local Examinations
Syndicate, Joint Matriculation Board, University of London School
Examinations Department, Oxford and Cambridge Schools Examination
Board, University of Oxford Delegacy of Local Examinations,
Southern Universities' Joint Board.*

*The Associated Examining Board, the University of London School
Examinations Department and the University of Oxford Delegacy of
Local Examinations accept no responsibility whatsoever for the
accuracy or method of working in the answers given in the text.
These are solely reflections of the opinion of the author.*

To my niece Helen

Acknowledgements

The author would like to thank two colleagues, Susan Jennings and Modris Bité, for helpful discussions during the preparation of this book. These colleagues and a student, Merryn Collard, supplied a number of the original drawings. Mr Geoffrey Goats of Rothamsted Experimental Station kindly provided background information on agriculture for Chapter 14.

Contents

Board Requirements
Key:

x = required
− = not required
(x) = limited knowledge
    required or not
    specifically mentioned
    in syllabus
o = optional

1. How to Use This Book

These Passnotes provide a step-by-step guide on how to study, revise and tackle examination questions in O-level Biology. Special attention is given to points the examiner is looking for, and common mistakes are pointed out. These Passnotes are not a textbook, but they do contain the basic information needed to gain a good O-level pass. They are best used alongside your class notes when you are revising. With proper use, they should enable you to enter the examination with confidence and achieve success.

O-level Biology examinations test a student's ability to:
1. Recall biological knowledge and express it in writing or in diagrams.
2. Recognize relationships between different areas of biology and, in particular, the relationship between structure and function.
3. Assess or interpret simple biological experiments and data.
4. Apply biological knowledge to new and unfamiliar situations to explain observations and to solve problems.

First of all, you need to know through which examination board you are taking your Biology examinations. The major English examination boards are listed in Table 1.1. Using the table you can find out how many papers you will be sitting and of what type – theory or practical. The table also shows which type of questions you will be asked on the theory papers. Within the table, fractions refer to parts of papers.

Next, you will need to know which topics you must learn for your syllabus. This information is given in the Contents list. The different examination boards are shown at the top and by reading down the right-hand side you will discover which topics you need to know. Some topics are optional and for others only a limited knowledge is required. This is indicated in the list. The major sections in the list correspond to chapters in the book, and each section is divided into numbered topics.

In your revision, you would be unwise to leave out major sections of the syllabus. A proper understanding of biological principles depends on you knowing how the different processes in an organism fit together to make a living, working unit. Miss out any process, and you do not have the overall picture. For example, how can you understand

Table 1.1. *G.C.E. O-level biology: examination papers by board*

BOARD	No. of papers	Short Answer	Multiple Choice	Longer Answer	Practical paper
AEB	2	1	–	1	–
CAM	3	$\frac{1}{2}$	1	$\frac{1}{2}$	1
JMB	1	$\frac{2}{5}$	$\frac{1}{5}$	$\frac{2}{3}$	–
LON	2	–	1	1	–
OXF	2	1	–	1	–
O&C	2	$\frac{2}{3}$	$\frac{1}{3}$	$1\frac{1}{3}$	$\frac{1}{3}$
SUJB	1	–	$\frac{2}{5}$	$\frac{3}{5}$	–

AEB　= Associated Examining Board
CAM　= University of Cambridge Local Examinations Syndicate
JMB　= Joint Matriculation Board
LON　= University of London School Examination Department
OXF　= Oxford Delegacy of Local Examinations
O&C　= Oxford and Cambridge Schools Examination Board
SUJB = Southern Universities' Joint Board

nutrition properly if you do not know about transport? After all, once food enters the body, it must be transported to where it is needed.

If, because of lack of time, you have to omit certain sections of the syllabus, you are probably wisest to leave section 13 till last, unless you find this section easy. However, you should consult your teacher about this.

By now you should know which examination board you are entered with, what type of examination papers you will be set, and you will be aware of which topics you need to know. It is worthwhile at this stage to get a copy of your board's syllabus and some copies of past examination papers. You can ask your teacher about this. With syllabus and papers in hand, you will be clear on the emphasis your board gives to different topics, and you will see for yourself the kind of questions you will be set, as well as the format of the examination papers.

You can use this book to refer to if you have any points of difficulty during your revision. Perhaps better, the Passnotes can be used alongside your class notes throughout your revision.

Chapters 3 to 15 contain the course contents and each chapter has the following layout:
1. Introduction.
2. The main body of the text, containing numbered topics, and including definitions and examples of examination questions.

3. Near the end of the chapter is a list of definitions (names only) and key words.

4. At the end of the chapter are a number of relevant examination questions from past papers.

The list of definitions and key words allows you to test yourself on the factual contents of the chapter. The examination questions test your understanding and give you the opportunity to plan examination answers.

The final chapter gives advice and encouragement on what to do just before, and during, your examination.

First of all, however, you should read Chapter 2. This gives a comprehensive guide on how to tackle the different types of examination question you will come across. Learning and revision methods are also included and suggestions are made on how to plan and organize your study.

2. *Working Towards the Exams*

Having worked out which topics you need to know for your syllabus, and what types of question you will be asked, you should then *plan* your study and revision. To do this, follow through the suggested sequence below, point by point.

1. First of all make sure your school or college notes are complete.

2. Tidy up your files or notebooks and make sure they include the following:

(a) clear underlined topic titles and sub-titles

(b) underlined 'key words'

(c) clear diagrams with titles

(d) a contents list, referring to numbered pages in your notes.

If you have any notes missing, or have any points or difficulty, see your teacher about this.

3. Several months before the exam, plan how much revision you need to do each week to finish revising your complete course *before* the exam. Break your coursework down into major sections and work out how many sections you need to revise each week or month in order to finish on time.

4. The use of compulsory short-answer and multiple-choice questions means that exam questions can be set covering the whole syllabus. You would do well to have a thorough knowledge of all parts of the syllabus. However, if you have particular trouble understanding certain isolated topics, e.g. Genetics (Chapter 13), you might be well advised to 'cut your losses' on some of these topics and concentrate on the rest of the syllabus. Again, you should see your teacher about this. Most of Chapters 3 to 12 is 'core' information which is necessary for a proper understanding of biological processes and will be required for you to give full and complete answers in Biology exams. To some extent you can be guided by the topics which have regularly come up in your board's exams in previous years.

2.1. *Revision Technique*

1. Revise the course topic by topic. You can use the section headings and sub-headings in this book to help you. Reduce your notes or chapters of this book into section and sub-section headings. Write these down on a separate sheet of paper. For example (see page 195):

Topic: Ear
Sub-topics: Hearing – outer ear
                          middle ear
                          inner ear
              Balance – utriculus
                             sacculus
                             semicircular canals

Read over each section and sub-section several times. Make a list of the key facts. Check your understanding of all the biological terms and definitions. At this stage it may be very helpful for you to construct a pattern diagram for each topic (see page 21).

You should be able to reach the stage where you can reproduce these sections and their contents *without* prompting from the book or your notes. The sub-headings and key words you remember should trigger off a mass of related facts in your memory.

2. Having reached the stage where you can recall topics by remembering section and sub-section headings and key words, do not let your efforts go to waste. Unless you *review* your notes regularly, you will rapidly forget them. Reviewing means quickly re-reading your notes and headings, and this should be done at one- or two-week intervals. This is very important. It is only by regular reviewing that your organized knowledge will be retained in your long-term memory.

3. It is sometimes difficult to remember the list of key facts which relate to a particular topic; for example, the functions of blood (page 135) or the functions of the liver (page 97). There are certain techniques to help you do this:

(a) One way is to use a **mnemonic**. These are words, sentences or phrases from everyday language which are simple to remember; these are linked to technical words which are difficult to recall. For example, the different regions of the vertebral column and the order in which they occur (see page 218), can be remembered using the following mnemonic:

Several	Cervical
Thirsty	Thoracic
Lumberjacks	**Lumbar**
were	
Stacking	**Sa**cral
Coal	Caudal

Wherever possible, make up your own mnemonic. Often ruder ones are easier to remember. Two more examples are found on pages 34 and 303.

(b) Another way to remember a long list is to group related items together and remember how many there are in each group. For example, the functions of blood (page 135) consist of 11 items which can be grouped into 4 categories:

1 function – formation of tissue fluid
5 transport functions
2 functions involving body temperature
3 functions involving defence against pathogens

(c) A very important way of remembering key words and their inter-relationships is to use pattern diagrams. Start off with the main points which relate to a topic, put these down on paper with plenty of space around them, and then gradually add key words, using arrows to show functional connections between them. For example, what do you know about the functions of the human small intestine (see page 90)? You can start off by writing down key words such as **duodenum**, **ileum**, **bile**, **pancreatic juice**, **digestion**, and **absorption**, and very quickly you can build up a pattern diagram linking information from your memory. For example see Fig. 2.1.

Research has shown that information is best memorized, understood and recalled when built up into pattern diagrams in the mind, rather like the parts of a jigsaw puzzle. When you recall several of the pieces in the puzzle, you can gradually fit the rest together. You will be surprised how much information you will be able to recall. Your own pattern diagram may not be like the one opposite; it will be unique to you. When you have finished revising a topic always try to summarize it in this way. When you come to the exams you will be able to remember your pattern diagrams and even create new ones when planning answers to essay questions.

4. Now, we will assume that all is going well and that you are keeping to your revision schedule, you are summarizing your notes as headings, sub-headings and key words, and that you are building up pattern diagrams for topics. The next step is to apply your knowledge and under-

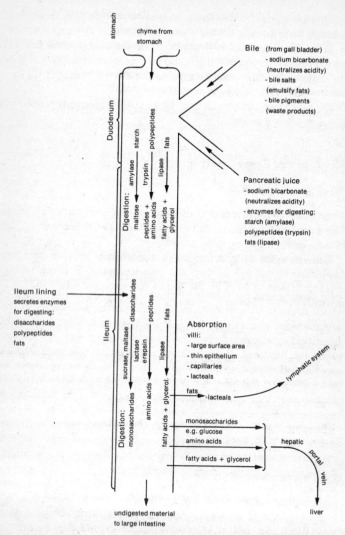

Fig. 2.1. *Pattern diagram for the human small intestine*

standing to answering examination questions. The questions in the text and at the end of each chapter are there to enable you to do this. To start with you may require prompting from your notes or this book to help you find the answers. Often, you will notice, the same question requires knowledge from different parts of the syllabus. You should reach the stage when you can answer questions without referring to your notes or to the book, and in the time allowed in the examination. Get your answers marked if you can; if not, use your notes and this book to check the answers. Multiple-choice questions are given throughout the book and answers are provided on page 346).

The next section in this chapter looks at the different types of exam question you will be set.

2.2. *Types of Exam Question*

Short-answer questions

These make up the first part or the whole of certain examination papers. Check in Table 1.1 (page 16) to see whether your board sets this type of question.

You are asked to give one-word, one-sentence or short-paragraph answers in spaces provided on the question paper. You can tell how much you need to write by the size of the space provided, and by the mark allocation. Here are two examples. Suitable answers are given.

Q. (a) Why do trees usually die when a complete ring of bark is removed from round the trunk?
The phloem is removed and this prevents organic substances such as sugars passing down the trunk to the roots. The roots eventually starve to death and cause death of the tree. [3]

  (b) Give one way in which this damage could happen naturally.
By deers eating the bark. [1]

[OXF]

Q. (a) Name *a hormone produced in the body of a mammal.* Insulin

  (b) Name *the organ where this hormone is produced.* Pancreas

  (c) Describe one *effect of this hormone in the body.*
    It stimulates the uptake of glucose by cells in the body and causes a lowering in blood glucose level. [3]

[AEB]

In other examples, you may be given a photograph, diagram, table or graph about which you have to answer several short questions. Keep your answers simple, brief, and to the point. Be guided by the size of space provided and by the mark allocation. In general, one mark is given for each fact. A definition or brief explanation may be worth two or more marks.

Multiple-choice questions

Multiple-choice questions, sometimes called fixed response or objective questions, make up all or part of certain examination papers. Check in Table 1.1 to see whether your board sets this type of question.

Normally, you are asked to choose the correct answer from a choice of four or five which are given. For example:

Q. $C_{12}H_{22}O_{11}$ is the chemical formula for
A. an amino acid
B. a protein
C. a fat
D. a carbohydrate
E. a nucleic acid.   [1]
                                                                   [LON]

If you do not know the correct answer right away, you can at least eliminate those which you know to be incorrect. For example, A, B and E can be discounted, because they all contain nitrogen. That leaves you with C and D, and if you blind guess, you have a 50 per cent chance of being right. You may, however, by this time, have thought out the correct answer, which is D (see page 44).

Students often find they do poorly on multiple-choice questions, although they expect to do well. Two common faults are misreading the question, or overlooking the best answer. Take great care with multiple-choice questions and double-check your answers. Sometimes the questions are rather lengthy and need to be read very carefully.

Usually multiple-choice questions ask for *one* correct answer. If you indicate two correct answers, you will automatically be marked wrong. If you do not know the answer to a question, you should choose an answer anyway. That way you at least have a chance of getting the question right and being awarded a mark.

On JMB and London Boards some of the questions are more complex than the type shown here. If you are entered for the examination with one of these boards, you should take the opportunity to look at and practise their multiple-choice questions *before* the exam.

Longer-answer questions

These make up the second part or the whole of an examination paper. There are two types of longer-answer question. The traditional essay-type question and the increasingly popular structured question.

Structured questions are made up of several parts, each of which usually requires a one- or two-paragraph answer. The length of the answer will be governed by the marks allocated to it. For example:

Q. Describe the chemical composition and biological functions of
(a) carbohydrates   [5]
(b) fats   [3]
   proteins   [5]
Give an account of the digestion of proteins in a mammal.   [12]
(Details of dentition, tooth structure and digestive organs are not *required.)*
                                                              [LON]

In longer-answer questions it is important to plan your answer before writing it out fully. There are a number of reasons for this. In marking your answer, the examiner is looking for a number of key facts, clearly explained and preferably in a logical order. Planning improves your ability to come up with a really clear and complete answer, free of irrelevant information.

The best way to answer a structured question is to look for the key words in the question, underline them, and plan your answer around these. For example:

The key words relating to parts (a), (b) and (c) of the above question are **chemical composition** and **biological functions**.

The answer plan for (a) is:

chemical composition ← carbohydrates → **biological function**
– contain C, H, O   – broken down to release
                                            energy (monosaccharides)
– 6-C sugar units   – energy stores (starch,
– monosaccharides   glycogen)
        ↓   – in plant cell wall (cellulose)
polysaccharides

Which can then be written out as:
(a) Carbohydrates are made up of the chemical elements carbon, hydrogen and oxygen combined to form 6-carbon sugar units. Monosaccharides, e.g. glucose, are made up of one of these sugar units, while polysaccharides are made up of many. Monosaccharides are broken down

in respiration to release energy. Polysaccharides such as starch (in plants) and glycogen (in animals) are used as energy stores. The polysaccharide cellulose is the structural material in the cell wall of plants.

Now try answering parts (b) and (c). The relevant information is given on pages 45 and 46. The key words in the second part of the question are **digestion**, **proteins** and **mammal**. Construct an answer plan using the information on page 94.

In a traditional essay-type question *you* will have to provide most of the key words and will have to construct a framework around these. For example:

Q. Describe the structure and functions of mammalian skin. [25]
[LON]

A diagram of mammalian skin in section is a good starting-off point (see Fig. 8.5, page 174).

The key words in the question are clearly **structure** and **functions**. Using these as your starting point, you should be able to list other key words under these headings.

Structure	Functions
Two layers:	Five main functions:
1. Epidermis	1. Protection
2. Dermis	2. Temperature regulation
	3. Sensitivity
	4. Excretion
	5. Vitamin D production

Now, each of these key words will lead on to others, and you will be building up your essay. See pages 173–5 and build up your essay plan using as a framework the key words above. You may be able to build up a pattern diagram showing structural and functional relationships (see page 21).

By providing a plan, you are not only ordering your thoughts and planning your essay logically, but you are helping yourself to remember the facts you know. Once you have written a fact down on paper, you can concentrate on remembering other facts. Also, without your outline to refer to, key points could easily be left out of your answer.

Part of your revision should consist of preparing essay plans or pattern diagrams for questions likely to come up in the exam. Get hold of past examination papers so that you can map out specimen answers.

Questions on experiments

A description of an experiment should be written up in a logical order under appropriate headings: 'Aim of the Experiment', 'Method', 'Results', 'Conclusion'. See, for example, the experiment described on pages 69–70.

For the 'Aim', do not write 'Demonstration of ...' or 'Experiment to prove ...', since this implies you knew the result before you did the experiment. Instead, follow proper scientific method and write 'Experiment to investigate ...' or 'Experiment to examine ...'.

In the 'Method' section, where appropriate, use a diagram of the apparatus to convey clearly as much information as possible. Your account can now follow: 'The apparatus was set up as shown in the above diagram.' In your written account do not duplicate information already given in the diagram. Concentrate on any special features of the apparatus and experimental procedure, such as how often the results were recorded and how long the experiment was run for. Include in your account or diagram, where appropriate, quantities used.

Write in the past tense and use the impersonal 'was recorded' not 'I recorded'. Do not forget to emphasize which was the **test** and which the **control** part of the experiment.

In the 'Results' section a plain statement of what happened in both **test** and **control** is required, without any discussion. If the rate of a process, or an amount, is being measured, use appropriate units.

The 'Conclusion' is the answer to the aim of the experiment. This is a plain statement, not an essay. In the question:

Q. Describe in detail an experiment which would demonstrate that light is needed for photosynthesis.   [13]

[LON]

the aim of the experiment would be 'to investigate the necessity for light in photosynthesis', and the experiment's conclusion is likely to have been 'light is necessary for photosynthesis'. You may have some comments or criticisms to make about the experiment. A brief mention may gain you some credit.

Diagrams

Drawing biological diagrams is a skill you should master. With patience, care and practice anyone can draw clear and accurate diagrams. They are a valuable source of marks in examinations.

Diagrams can be used in longer-answer questions whenever they make your answer clearer and save time. Alternatively, diagrams may be specifically asked for in questions, for example, 'Give an illustrated account of . . .' or 'With the help of diagrams describe . . .'.

In longer-answer questions diagrams can be used to complement written text. In the text you can freely refer to your diagram, but do not simply repeat information shown in the drawing. Remember, diagrams are supposed to make your answer clearer and save time.

Diagrams are simplified, explanatory drawings. They are not the same as the 'true-to-life' drawings done in practical classes or practical examinations (see page 32).

When drawing diagrams, take note of the following points:

1. Draw in pencil (that way you can easily rub out mistakes).
2. Your diagram should be large (at least twice the size of those shown in this book).
3. The diagram should be made up of distinct, continuous lines (do not sketch).
4. The diagram should be fully labelled.
5. Label in ink.
6. A straight line should run from the label to the structure to which it refers (do not use curved or branched lines).
7. Label lines should not cross.
8. The diagram should have a specific title.

We can apply these points in answering an examination question:

Q. Make a labelled diagram to show the arrangement of structures in the leaf of a flowering plant as seen in vertical section. Indicate on your diagram the distribution of chloroplasts. [9, 3]

[OXF]

Fig. 5.2 on page 66 is a suitable answer to this question. The upper diagram is a general plan showing the distribution of tissues in the leaf. The lower diagram is an expanded detail of part of the leaf. This shows the shape, sizes and arrangement of cells in the different tissues. Drawing many identical structures in a diagram is a waste of time. It is best to draw carefully a few of the structures and indicate that many more are present. The question asks you to indicate the distribution of chloroplasts, so obviously some detail of cell contents is required. In the example shown, only one cell from each tissue type is drawn in detail.

In drawing, use a sharp pencil – a 1H or HB pencil is best. First, map out the general shape lightly in pencil. Divide up the diagram into its major parts, making sure the proportions are correct. Fill in any detail last of all. There is nothing worse than starting at one end of your diagram, drawing in great detail, and getting to the other end and finding it will not fit. Always begin with a general plan, getting the proportions of the major parts right, and then fill in the detail last.

It is well worth copying out all the major diagrams from your class notes or from this book. Then try drawing them *without* referring to the original. Be critical when assessing your diagrams. Are they clear, accurate, properly labelled and titled?

In general, you should avoid colouring or shading. This takes time and may make your diagram less clear. In diagrams of the mammalian circulatory system, blood vessels which carry oxygenated blood can be shown by a red line, and those carrying deoxygenated blood by a blue line. If you do this, you should provide an explanatory key.

Always give a title to your drawing, and be as specific as possible. For example:

Q. Make a labelled diagram to show the structure of a mammalian neurone.
[6]
[OXF]

There are three main types of neurone (see page 182) and whichever one you draw, you should give its appropriate name. If you miss this out you may be penalized heavily.

When drawing a section, always give its orientation. A transverse section (T.S.) is really another name for a cross-section. In a structure which has a long and a short axis, a T.S. is a cut across the long axis. A longitudinal section (L.S.) is a cut along the long axis (see Fig. 2.2).

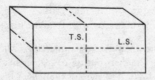

Fig. 2.2. *Orientation of sections*
T.S. = transverse section
L.S. = longitudinal section
V.S. = vertical section

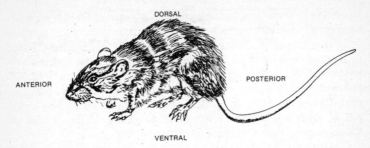

Fig. 2.3. *Orientation of the body*

A vertical section (V.S.) is an upright section and is commonly used for describing sections through the kidney, heart and skin. It is particularly used for structures which do not have a noticeable long or short axis.

In animal drawings you should be aware which is the anterior (head) end, the posterior (tail) end, and the front (ventral) and back (dorsal) surface – see Fig. 2.3.

Graphs

You may have to construct a graph from data given to you, or you may be provided with a graph, and have to answer questions by interpreting or reading off the graph.

In drawing graphs, take note of the following points:
1. Plan how you are going to fit the graph on to the paper provided.
2. The horizontal or x-axis is used for the controlled variable. This means those units which are determined and controlled by the experimenter.
3. The vertical or y-axis is used for the uncontrolled variable. These are the results which the investigator finds out from the experiment.
4. Each axis must be divided into the correct number of units marked off at equal intervals.
5. Each axis must be fully and accurately labelled.
6. Plot each point with a circled dot or with a small cross.
7. Join up the points with a smooth curve or with straight lines drawn with a ruler.
8. The graph should have a title.

Let us try this out on an examination question.

Q. Twenty tadpoles were put into each of two aquaria on day 0. The water in one aquarium was kept at 28 °C and the water in the other at 7 °C. At intervals the lengths of all the tadpoles were measured and the average length of each batch calculated to give the following results:

day	average length in mm	
	tadpoles kept at 28 °C	tadpoles kept at 7 °C
0	*15*	*15*
1	*19*	*17*
6	*25*	*19*
9	*29*	*20*
13	*33*	*21*
16	*35*	*21*
20	*38*	*22*
23	*30*	*22*

(a) *Plot these results on the squared paper using one pair of axes.*

[6]

[OXF]

What we have here are two curves plotted on one set of axes.

Time (in days) is the controlled variable – it is determined by the experimenter. It goes on the horizontal axis. The average length of tadpoles (in mm) is the uncontrolled variable – it is what the experimenter is finding out. It goes on the vertical axis.

Now we can plot the graph as follows (for the time being ignore the arrowed broken lines):

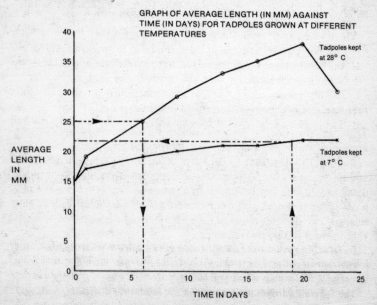

GRAPH OF AVERAGE LENGTH (IN MM) AGAINST TIME (IN DAYS) FOR TADPOLES GROWN AT DIFFERENT TEMPERATURES

Tadpoles kept at 28° C

Tadpoles kept at 7° C

AVERAGE LENGTH IN MM

TIME IN DAYS

The question then goes on, with further sections asking us to interpret from the graph.

(*b*) *On day 19 what was the average length in mm of the tadpoles kept at 7°C?* [1]

To do this, we find day 19 on the horizontal axis, draw a vertical line leading up to the appropriate curve, and then draw a horizontal line leading to the vertical axis. Here we read off the value, which is 21·7 mm.

We can easily do this the other way round, working from a value on the vertical axis to one on the horizontal axis. For example, if we are given the average length of tadpoles grown at a particular temperature, we can find out how long they have been in the aquarium. The tadpoles grown at 28°C, and which have an average length of 25 mm, have been in the aquarium 6 days.

(*c*) *In any one day what was the greatest average growth in length shown by any of the tadpoles?* [1]

In a graph, the slope of the curve indicates the rate of change. The steeper the slope, the faster the rate of change.

In our graph, we are looking for the part of the curve which has the steepest uphill slope. This will give us the greatest average growth in length per day shown by any of the tadpoles. This slope is found between day 0 and day 1 for tadpoles grown at 28°C. Here, the tadpoles increase in length by 4 mm in one day. The answer to the question is 4 mm.

(*d*) *On which day was there the greatest difference in average length between the two batches?* [1]

Day 20 is the answer. Here the two curves are farthest apart.

Practical questions (O&C and CAM only)

These are designed to test your ability to follow instructions, make accurate observations and come to valid conclusions.

Follow all instructions to the letter. If asked to examine a specimen and record your observations, take the opportunity to include as much detail as you can, but be guided by the mark allocation. For example, if examining a flower, you can include detail of size, shape, colour and smell. *Marks are only awarded for what you write on the paper*, so keep your answers precise and detailed.

If asked to make a drawing of a specimen, draw what *you* see, not a textbook drawing. Use pencil, draw large and fully label and title your drawing; include scale or magnification.

The following topics are popular in exams: flower structure, fruit and seed structure, vegetative organs (e.g. bulbs, corms and rhizomes, twigs), comparisons/drawings of bones or teeth, external features of small invertebrates, stages in life cycle of insects. Simple experiments may be set on topics such as osmosis, behaviour in small invertebrates, or using food tests.

3. The Organization of Life

To begin to appreciate what a living organism is, in a biological sense, you need to know what it is made of and what it is capable of doing. Despite the enormous range of life forms – from microscopic specks to oak trees and human beings – all living organisms are made up of similar substances and show certain characteristic features.

3.1. The Chemistry of Life

Organisms are made up of four common chemical elements – hydrogen (H), carbon (C), nitrogen (N) and oxygen (O). Seven other elements (sodium (Na), magnesium (Mg), phosphorus (P), sulphur (S), chlorine (Cl), potassium (K), and calcium (Ca)) are present in moderate amounts. A number of elements, such as iron (Fe) and iodine (I), are also present in minute amounts.

These elements are arranged into inorganic and organic molecules. **Inorganic molecules** tend to be small and simple and consist of two or more elements chemically bonded together. Inorganic molecules are abundant in both living and non-living things. One of the commonest is water (H_2O).

Surprisingly enough, most organisms are made up of over two-thirds water. The chemistry of life is dominated by the chemistry of water.

1. Water is the **universal solvent**. It is the liquid in which substances in the body are dissolved, and all the body's chemical reactions take place in it.

2. Water itself takes part in certain chemical reactions, such as photosynthesis (page 64) and digestion (page 82).

3. It serves as a medium for the transport of materials (page 132).

Organic molecules are a characteristic feature of living, or once-living, organisms. These molecules are based on carbon and hydrogen to which oxygen and other elements are often chemically bonded. Such molecules tend to be complex and sometimes very large. In living organisms, four types are important: carbohydrates, proteins, fats and nucleic acids. These are dealt with in section 3.8.

3.2. *The Characteristics of Life*

All living organisms demonstrate *all* seven of the following characteristics at some time in their life.

1. **Movement**: all living things move either a part or the whole of their bodies (Chapter 10).

2. **Feeding**: living things feed. Food is the material from which organisms obtain (i) the energy, and (ii), the raw material, to construct, repair and maintain the body. The study of food and feeding processes is termed **nutrition** (Chapter 5).

3. **Respiration** is a chemical reaction, or rather a series of chemical reactions, by which organisms release energy from organic foods (Chapter 6).

4. **Excretion** is the removal of waste substances produced by chemical reactions in the body (Chapter 8).

5. **Growth**: living organisms grow, i.e. increase in size. The term growth is usually taken to include increase in complexity as well as size (Chapter 12).

6. **Reproduction** is the production of new individuals (offspring) derived from existing individuals (parents) (Chapter 11).

7. **Sensitivity**: living organisms are sensitive. They have the ability to respond to a stimulus (Chapter 9).

A convenient way to remember these seven characteristics is to use a mnemonic such as:

Many	Movement
Foolish	Feeding (nutrition)
Rabbits	Respiration
Eat	Excretion
Green	Growth
Rhubarb	Reproduction
Shoots	Sensitivity

These features are not necessarily very easy to demonstrate. In man, for example, respiration is accompanied by obvious breathing movements (see page 123). Plants, however, do not breathe and complex apparatus is required to demonstrate respiration (see page 113).

Non-living objects may show certain characteristics of living organisms, but not all of them. A motor car takes in oxygen and gives out carbon dioxide (respires), it consumes fuel (feeds), it releases exhaust fumes (excretes), and it moves. Arguably, it is also sensitive (in a biological sense). If you switch on the ignition the car will (if you are lucky) respond by the engine starting. What a car does not do, however, is grow or reproduce. It does not fulfil all the seven characteristics of life.

3.3. *Metabolism*

Living things are complex arrangements of substances not normally found in non-living situations. To make complex substances from simpler ones requires energy. Organisms thus need a continual supply of energy to make the organic substances from which they are constructed. The energy is used in forming chemical bonds within these molecules.

Metabolism is a word used to describe all the chemical reactions which occur in an organism and are necessary for life. There are two types of metabolism: **catabolism** and **anabolism**.

Catabolism is the process by which complex substances are broken down into simpler ones, resulting in the release of energy. Respiration (page 105) is an important example of catabolism.

Anabolism is the process by which simple substances are built up into complex compounds. This type of reaction requires a supply of energy. Examples of anabolism are photosynthesis (page 64) and all cases of growth and repair in the bodies of organisms.

Both catabolic (breaking down) and anabolic (building up) processes occur in living things. Energy released from catabolic reactions is used to drive anabolic reactions.

3.4. *Cells*

Organisms are made up of cells, though there are some exceptions to this (page 43). A cell consists of a **nucleus** suspended in **cytoplasm** and enclosed in a **cell** (plasma) **membrane**. The living parts of the cell – the nucleus, cytoplasm and cell membrane – are sometimes called **protoplasm**.

Definition: The cell is the basic unit of life. It is a unit of protoplasm made up of a nucleus and cytoplasm surrounded by a cell membrane.

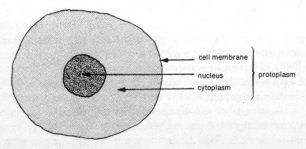

Fig. 3.1. *The chief characteristics of cells*

Some organisms, such as *Amoeba* (page 326) and *Chlamydomonas* (page 310) are made up of one cell and are called single-celled or **unicellular**. Most organisms are many-celled or **multicellular**. Oak trees and human beings, for example, contain billions of cells.

Living cells are small and there are good reasons for this. One reason is that all the living parts of the cell must be in close contact with their surroundings to obtain supplies and get rid of wastes quickly enough to support life. Movement of substances into and out of the cell is mainly by diffusion (page 55) and this will only occur rapidly across short distances. Hence, cells are usually microscopic (visible only with a light microscope). The largest cells, e.g. bird's eggs (page 342) and larger plant cells, contain a large amount of non-living matter.

Cells come in different shapes and sizes and the appearance of a cell (its structure) is closely linked to the function it performs. For example, a nerve cell (page 182) is usually long and thin and transmits nerve impulses from one precise point to another in an animal's body. The cell has a fixed shape and does not move around inside the body. In contrast, certain types of white blood cell (page 134) can move around and change shape. They are **phagocytic**, meaning they can engulf particles by flowing round them. Their function is to mop up foreign organisms which get inside the body. They can feed themselves, whereas the nerve cell requires its food delivered.

This relationship between structure and function is a very important theme in biology. We will return to it again and again. At all levels of organization (see Fig. 3.2), from biological molecule to complete organism, structure is related to function.

3.5. *Plants and Animals*

Living organisms are usually separated into two major groups (kingdoms), plants and animals (see page 303). These groups show the following differences (based on flowering plants and mammals):

1. **Cell structure**: plant cells have a large vacuole and a cell wall made of cellulose. Animal cells have neither of these. Plant cells may contain chlorophyll, animal cells never do.

2. **Feeding**: green plants (plants which contain the green substance **chlorophyll**) are **autotrophic**, that is they make their own organic food. They do so by the process of **photosynthesis** (page 64), using chlorophyll to trap sunlight energy. Animals are **heterotrophic** (page 63), that is, they get their organic food 'ready-made'.

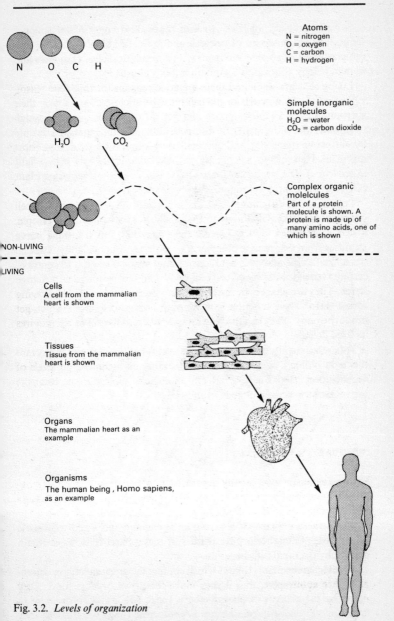

Atoms
N = nitrogen
O = oxygen
C = carbon
H = hydrogen

N   O   C   H

Simple inorganic
molecules
H_2O = water
CO_2 = carbon dioxide

H_2O CO_2

Complex organic
molelcules
Part of a protein
molecule is shown. A
protein is made up of
many amino acids, one of
which is shown

NON-LIVING

LIVING

Cells
A cell from the mammalian
heart is shown

Tissues
Tissue from the mammalian
heart is shown

Organs
The mammalian heart as an
example

Organisms
The human being , Homo sapiens,
as an example

Fig. 3.2. *Levels of organization*

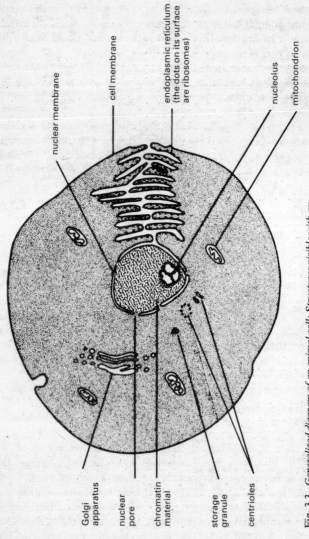

Fig. 3.3. *Generalized diagram of an animal cell: Structures visible with an electron microscope*

nuclear membrane

cell membrane

endoplasmic reticulum (the dots on its surface are ribosomes)

nucleolus

mitochondrion

Golgi apparatus

nuclear pore

chromatin material

storage granule

centrioles

3. **Movement**: movement is faster and more obvious in animals than in plants. Most animals can move their bodies from one place to another, i.e. they show **locomotion**. Plants tend to move slowly and usually do so using only part of the body.

4. **Sensitivity**: both animals and plants are sensitive (respond to stimuli) but plant responses tend to be much slower. In plants, responses are controlled by hormones; in animals, by the nervous system as well as hormones.

5. **Growth**: in plants, growth is usually restricted to certain regions called **meristems** (page 262). Growth is **unlimited**, meaning it continues throughout life. In animals, growth occurs throughout the body and is **limited**; growth slows down and stops when adult size is reached.

3.6. *Cell Structure*

When viewed through a light microscope (which magnifies up to 1,500 times), only a few distinct structures are visible within the cell. Staining helps to reveal the nucleus, grainy cytoplasm and cell membrane. When examined with an electron microscope (which magnifies up to 500,000 times) many more structures became visible (see Fig. 3.3). The cytoplasm ceases to be a grainy mass, but is seen to be made up of compartments. These compartments, together with the nucleus, are called **organelles**. They have different shapes and sizes and perform different functions.

Definition: an organelle *is a structure within the cell which has a special function or functions, e.g. the nucleus.*

The following features are common to all cells (those marked with an asterisk (*) are visible only with an electron microscope):

Cytoplasm: a grainy, jelly-like material forming the groundwork of the cell.

Plasma (cell) **membrane**: a thin semi-permeable membrane (page 57) enclosing the cytoplasm. It is composed of protein and fat (page 45) and controls the passage of substances in and out of the cell.

Nucleus: the nucleus, bounded by the **nuclear membrane**, controls the activities of the cell (see page 49). It contains threadlike **chromosomes** which carry instructions for building the cell and controlling cell function. The **hereditary material** within the chromosome carries the instructions for making a new cell or organism. Chromosomes are not visible except during cell division (page 272). At other times they are dispersed throughout the nucleus as darkly staining **chromatin**. The nuclear membrane has pores to allow the exchange of substances between the nucleus and cytoplasm.

Mitochondria: these sausage-shaped organelles are the sites of respiration (energy release) in the cell. They have a folded inner membrane giving a large surface area for the attachment of respiratory enzymes.

Endoplasmic reticulum* (E.R.): this is a network of folded membranes often found throughout the cytoplasm. It is thought to transport substances (particularly proteins) around the cell. On its surface are often found **ribosomes**.

Ribosomes*: these are the sites of protein synthesis (page 49). They are usually found on the surface of the E.R. but may be scattered throughout the cytoplasm.

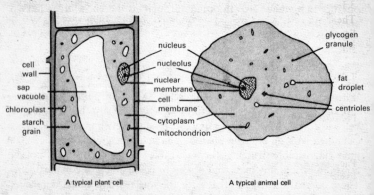

Fig. 3.4. *Comparison between a plant and an animal cell*

The following features are present only in plant cells (see Fig. 3.4.):

Cell wall: a rigid wall made of cellulose and surrounding the cell membrane. It gives shape and some support to the cell.

Vacuole: a large, non-living, fluid-filled space in the cytoplasm. It pushes the cytoplasm out against the cell wall and makes the cell turgid (page 60).

Chloroplasts: similar in size to mitochondria, these are the sites of photosynthesis in plant cells. They contain the green pigment **chlorophyll** together with the enzymes needed for the reactions of photosynthesis. (note: not all plant cells contain chloroplasts).

Starch: this is the storage carbohydrate in plant cells. It is found as grains in the cytoplasm.

The following features are found only in animal cells (see Fig. 3.4):

Centrioles*: during cell division, these cylindrical structures give rise to spindle fibres.

Glycogen: in animal cells glycogen is the storage carbohydrate. It is found as granules in the cytoplasm.

3.7. *Cells, Tissues and Organs*

All larger animals and plants are multicellular (many-celled). In these organisms the cells are grouped together to form tissues and organs.

Tissues are groups of cells with a similar appearance and function. Some tissues have just one type of cell, but most have two or three types mixed together.

Definition: a tissue *is a collection of similar cells performing the same function, e.g. muscle.*

The main types of tissue in plants and animals are listed in Tables 3.1 and 3.2. Notice that the tissues are classified according to function. There are other ways of classifying them. Rather than simply learn the list, you should look up the relevant page numbers. For each tissue consider how its structure is related to its function.

Table 3.1. *Plant tissues* (based on the flowering plant)

Name of Tissue	Functions	Examples
Protective tissue	Protection (also against desiccation)	Epidermis (page 320) Cork – dead cells (page 320)
Packing (supporting) tissue	Fills in space between other tissues and cell turgidity provides support	Parenchyma (page 229)
Photosynthetic tissue	Photosynthesis	Mesophyll (page 65)
Vascular tissue	Transporting water and food substances	Xylem – dead cells (page 149); phloem (page 149)
Strengthening tissue	Supporting the plant	Collenchyma; sclerenchyma – dead cells (page 229)
Meristematic (dividing) tissue	Cell division. Gives rise to other tissue types and enables growth	Shoot and root tips; cambium (page 320)

Organs are groups of tissues which combine together to form a structure which has a particular function or functions. An example is the mammalian heart (page 141). It contains epithelial tissue, cardiac muscle, nervous tissue and blood, all organized into a complex structure. One of the heart's functions is to pump blood around the body.

Definition: an organ *is a structure formed by two or more tissues working together to perform particular functions, e.g. the heart.*

Table 3.2. *Animal tissues* (based on the human)

Name of Tissue	Functions	Examples
Epithelial tissue	Lines and protects tubes and cavities within the body. Also protects body surface	Epidermis of skin (page 173); lining layer of the gut (page 91)
Connective tissue: has a matrix (non-living part)	Connects, binds, supports and protects (blood and lymph are exceptions)	Fibrous connective tissues such as tendons and ligaments (page 215); skeletal tissues – bone and cartilage (page 215); blood – a fluid tissue (page 135); lymph – a fluid tissue (page 146)
Muscular tissue	Brings about movement	Smooth (visceral) muscle; striped (skeletal) muscle; cardiac (heart) muscle (page 222)
Nervous tissue	Communication – transmits nerve impulses	Conducting tissue: motor, sensory and relay (page 182); non-conducting tissue: myelin sheath

In flowering plants (page 321) there are four major organs – the stem, root, leaves and, at certain times, the flower. The leaf, for example, is made up of epidermal, photosynthetic, packing, vascular and strengthening tissues (see pages 41 and 66). The leaf's main function is photosynthesis (page 64).

In the more complex animals such as insects (page 335) and mammals (page 342) the organs do not operate independently but work together as **organ systems** to take on the responsibility for particular functions. For example, in the human, the gut, liver and pancreas work together to digest and absorb food. They form the digestive system (page 83).

Figure 3.2 (page 37) attempts to show the relationship between chemical substances, cells, tissues and organs, and the organism as a whole. Clearly, not all organisms show all these levels of organization. Unicellular organisms, such as *Amoeba* (page 326) and *Chlamydomonas* (page 310), are not made up of tissues or organs; all life processes are carried out within a single cell. Some simpler multicellular organisms such as the coelenterate, *Hydra* (page 328), function at the tissue level of organization; life processes are carried out mainly by tissues and isolated cells.

A number of organisms are not made up of cells as we defined them earlier:

1. The non-green plants, **fungi** (page 314), have tubular structure called hyphae which contain cytoplasm. Within these are found many nuclei. No membrane or cell wall separates off one nucleus and region of cytoplasm from another to form recognizable cells.

2. **Bacteria** (page 307) are unicellular but do not have a proper nucleus. The genetic material is free in the cytoplasm and is not enclosed in a nuclear membrane.

3. **Viruses** (page 305) are on the borderline between living and non-living. They are extremely small (visible only with an electron microscope) and have a structure unlike any other kind of organism. They have no plasma membrane, nucleus or 'cytoplasm'.

3.8. *Biological Molecules**

You need to know the structure and function of the major types of biological molecule. Without this, you will not properly appreciate biological processes such as nutrition, respiration or excretion. In addition, several short-answer questions are usually set on the structure and function of biological molecules. Such knowledge is assumed in longer-answer questions.

Q. What are carbohydrates? [6]
[LON]

Q. Chemically, what is fat? [2]
[OXF]

Carbohydrates

These contain the elements carbon, hydrogen and oxygen.

Carbohydrates are made up of units called six-carbon sugars. These units can be joined together or broken apart, rather like beads on a string. **Monosaccharides** are made up of one unit, **disaccharides** are made of two, and **polysaccharides** are made up of many.

---

* Check with your syllabus how much detail is required.

	Chemical formula	Solubility	Examples
Simple sugars (monosaccharides) one sugar unit o	$C_6H_{12}O_6$	Soluble	Glucose Fructose Galactose
More complex sugars (disaccharides): two sugar units o—o	$C_{12}H_{22}O_{11}$	Soluble	Maltose (milk sugar) Sucrose (table sugar)
Large molecule carbohydrates (polysaccharides): many sugar units o—o—o—o—o—o—o	$(C_{12}H_{22}O_{11})_n$	Insoluble	Starch (in plants) Cellulose (in plants) Glycogen (in animals)

Soluble carbohydrates are called sugars. In both animals and plants, simple sugars can be built up into complex carbohydrates and then broken down again as required. Carbohydrates are transported round the body in solution as sugars and are stored within cells as complex insoluble molecules. Being insoluble, these large storage molecules do not produce an osmotic potential (see page 59) and so do not create problems of water movement. The joining together of sugar units is a **condensation** reaction. A molecule of water is released when each chemical bond between two sugar units is formed. The breaking apart of sugar units is a **hydrolysis** reaction. A molecule of water is used in breaking each bond between units. Digestion (page 82) is a hydrolysis reaction.

	Condensation	*Hydrolysis* (the reverse of condensation)
monosaccharide	o ↓	o ↑
disaccharide	o—o ↓ ⟶ H_2O	o—o ↑ ⟵ H_2O
polysaccharide	o—o—o—o—o — — — ↓ ↓ H_2O H_2O	o—o—o—o—o — — — ↑ ↑ H_2O H_2O

The building up and breaking down of carbohydrates

Functions

1. Monosaccharides are the main substances broken down in respiration (page 105) to release energy.
2. Polysaccharides are used as energy stores – in plants, **starch**; in animals, **glycogen**.
3. The polysaccharide **cellulose** is used by plants as structural material in the cell wall.

Proteins

Proteins contain the elements carbon, hydrogen, oxygen, nitrogen and often sulphur.

In the same way that polysaccharides are made up of chains of sugar units, so proteins are made up of chains of amino acids. However, whereas there is a choice of only one or a few different types of sugar unit in a carbohydrate, in a protein there are up to 20 different types of amino acid. These can be joined together in any order to form chains hundreds of units long, called **polypeptides**. The arrangement, or sequence, of amino acids in the chain varies from one protein to another. For example, in one protein

and in another

where each symbol represents a different amino acid.

Just as with polysaccharides, proteins can be broken down by hydrolysis into their constituent units (amino acids) and then built up again by condensation. They are transported round the body as the single units and are built up into chains as required by cells.

Functions

1. Proteins make up most of the structural parts of the cell and so are needed in the growth, repair and maintenance of all living cells.
2. Enzymes, the chemicals which regulate cell chemistry (page 49), are made of protein.
3. Antibodies (page 148) and certain hormones (page 181) are made of protein.

The immense variety of life is largely due to the enormous number of different proteins that exist.

Fats

Fats, like carbohydrates, contain only carbon, hydrogen and oxygen. However, in fats, the proportion of oxygen is lower. A typical fat is tristearin, $C_{57}H_{110}O_6$.

The most common type of fat has four parts, one part derived from glycerol and three parts from fatty acids.

A fat molecule

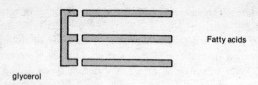

glycerol   Fatty acids

When a fat molecule is broken down by hydrolysis, the glycerol part is separated from the three fatty acids. The original fat, or a different one, can be built up by condensation.

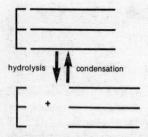

hydrolysis   condensation

Functions
1. They form part of the membranes of cells.
2. They are an important energy store and in respiration (page 105) release twice as much energy per gram as carbohydrates or proteins.
3. In mammals, fat stores protect internal organs from damage and fat under the skin acts as heat insulation.

Carbohydrates, proteins and **fats** are classed as foods (see page 82). Their presence in organic material can be determined using food tests (see Table 3.3). Questions on these tests are extremely popular in exams.

Table 3.3. *Food tests*

Food	Test	Positive result
1. CARBOHYDRATES		
(a) *Reducing sugars*: monosaccharides (simple sugars) and the disaccharide maltose	Make a solution of the test substance in water. Add an equal quantity of **Benedict's Reagent**. Gently bring to the boil	An orange or green precipitate
(b) *Non-reducing sugars*: disaccharides other than maltose (NOTE: this test should only be used after test (a) has given a negative result)	Dilute hydrochloric acid is added to the test sample and the resulting solution boiled. If a non-reducing sugar is present it will be hydrolysed to a reducing sugar. The solution is then neutralized with sodium carbonate and the test for a reducing sugar carried out as before	An orange or green precipitate
(c) *Polysaccharides*	Add iodine solution to the test substance	For starch a blue-black colour forms; for cellulose: blue; for glycogen: reddish-brown
2. PROTEINS	Use either of these tests:	
	1. Place test substance in a test tube. Add twice the volume of **Millon's Reagent** and heat gently	The test substance turns red on warming
	2. **Biuret test** for dissolved proteins. To the test solution add an equal volume of sodium hydroxide and one or two drops of copper sulphate solution	On shaking, a mauve coloration appears

Food	Test	Positive result
3. FATS AND OILS	Use any of the following tests:	
	1. Rub test solid on to filter paper	Grease spot forms on paper
	2. Shake test substance with ethanol. Pour ethanol into water	Milky suspension
	3. Add few drops of red dye Sudan III to the test substance in water. Shake. Allow to stand	Red oil layer floats on surface

Q. For each of the following food materials describe one test you would use to show its presence:

(i)   a reducing sugar;

(ii) starch;

(iii) a protein;

(iv) a fat.

[15]

[OXF]

Nucleic acids

These are made up of the elements carbon, hydrogen, oxygen, nitrogen and phosphorus. They are large, complex molecules made up of units called nucleotides joined together in long chains. Their function is the storage and translation of genetic information (page 271).

There are two types of nucleic acid: deoxyribonucleic acid(**DNA**) and ribonucleic acid (**RNA**). DNA is found in the chromosomes (page 271) within the nucleus of the cell. DNA carries a coded message for manufacturing proteins. The message is carried from the nucleus to the cytoplasm by RNA. At the ribosomes (page 40) within the cytoplasm proteins are made under instructions from the RNA.

Both the structural parts of the cell and the enzymes which control cell metabolism are made of protein. Thus the functioning of a cell is determined by the proteins it contains. The DNA, by controlling which proteins are made in a cell, controls the functioning of the cell. This is what we mean when we say the nucleus controls the cell's activities.

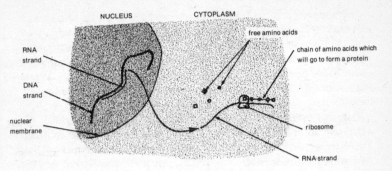

Fig. 3.5. *How the nucleus controls the cell (see text)*

3.9. *Enzymes*

Enzymes are 'biological catalysts'. Catalysts are chemicals which speed up chemical reactions. The catalyst itself remains unchanged by the reaction.

Unlike the inorganic catalysts used in the chemical industry, enzymes are made of protein. Without enzymes, chemical reactions in the body would be too slow to support life.

Enzymes, just like other proteins, are manufactured within cells. The properties of enzymes can be summarized as follows:

1. They are proteins made by living cells (they sometimes incorporate vitamins).
2. They are catalysts.
3. They are specific, i.e. one type of enzyme catalyses one type of reaction.
4. An enzyme works best at a particular temperature (the optimum temperature). As with other proteins, they are destroyed (denatured) by heating.
5. Each enzyme works best at a certain degree of acidity or alkalinity (the optimum pH).

Outside its optimum temperature and pH an enzyme may cease to work or may even be destroyed (we normally say '**denatured**' to indicate its chemical nature has been lost). The optimum temperature for most enzymes is the human body temperature, 37 °C. The optimum pH for enzymes varies. For pepsin (page 89) it is pH 2 (very acid), and for salivary amylase (page 89) pH 7·5 (nearly neutral).

Often the name of an enzyme ends in **-ase**. The first part of the name may indicate the substrate (reactant) whose reaction the enzyme catalyses. For example, **proteases** catalyse the breakdown of proteins; **carbohydrases**, the breakdown of carbohydrates; and **lipases**, the breakdown of lipids. Individual enzymes, however, often do not follow this rule. **Pepsin** (page 89) is an example of a protease. **Salivary amylase** (page 89) is a carbohydrase which hydrolyses (digests) starch.

Enzymes catalyse *all* chemical reactions in the body. However, the enzymes you are probably most familiar with are digestive enzymes. These are secreted into the gut and are, in fact, unusual enzymes because they are **extracellular** (work outside cells). Most enzymes are not secreted but are **intracellular** (work inside cells). One such enzyme in plants is **starch phosphorylase** found in the cotyledons of bean seeds (page 247). It catalyses the first reaction in the conversion of starch to sugar and is therefore important in germination, when stored carbohydrates are being utilized. An example of an intracellular enzyme in animals is **carbonic anhydrase** (page 133) found in the red blood cells of mammals.

Questions on the effect of temperature and pH on enzyme activity, together with experiments, are very popular with examiners. For example:

Q. (a) What are the effects of a change in
(i)   temperature
(ii) pH upon the rate of action of any named *enzyme*   [7]
(b) Describe with full experimental details how you would investigate either
of these effects.   [12]

[OXF]

In such questions it is usual to use a digestive enzyme as the named example, in the experiment, and to use a food test (page 47) to test for the presence of substrate or product.

Investigating the effect of temperature on the digestion of starch by salivary amylase (*ptyalin*)

This experiment investigates how rapidly starch is digested at different temperatures under the action of salivary amylase. When a starch solution no longer gives a positive test with iodine (see page 47) it can be assumed that the starch has been digested.

Method

1. A 1 per cent starch solution is made up and $5\,cm^3$ added to each of five test tubes.
2. $1\,cm^3$ of freshly collected saliva is added to each of four other test tubes.
3. The test tubes are placed in water baths at various temperatures as shown.

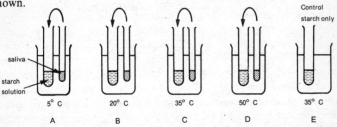

Fig. 3.6. *Apparatus for investigating the effect of temperature on starch digestion*

4. The test tubes are left in their respective water baths to reach the required temperature, and then the solutions in the two tubes are mixed.
5. At timed one-minute intervals a drop of each mixture is removed by dropper pipette and tested for starch, using iodine. NOTE: iodine is *not* added direct to the mixture – this would affect the reaction.
6. For each mixture, the time is recorded when a drop of the solution no longer turns the iodine blue-black, i.e. starch has been digested. The experiment is stopped after 20 minutes.

Results

Typical results are as follows:
   A. Starch test still positive after 20 minutes.
   B. Starch test negative after 8 minutes.
   C. Starch test negative after 2 minutes.
   D. Starch test still positive after 20 minutes.
   E. Starch test still positive after 20 minutes.

If mixtures A and D are then transferred to 35 °C for 20 minutes, A will give a negative test for starch and D a positive test.

Conclusions

The results taken as a whole indicate that the enzyme works best (digests starch fastest) at 35 °C, and at a lower temperature the reaction is slower. At 50 °C, the enzyme is denatured. This is indicated by the enzyme not working when kept at 35 °C. The control shows that starch does not break down of its own accord.

Investigating the effect of pH on the digestion of albumin by pepsin

This experiment investigates how rapidly cooked egg albumin (a protein) is digested at various pH conditions under the action of pepsin.

Method

1. Into each of 6 tubes is put a 5 mm cube of cooked egg white and to each tube is added a thymol crystal to prevent bacteria digesting the starch.
2. $2 \, cm^3$ of water, or $0.1M$ solutions of sodium carbonate or hydrochloric acid, are added to the tubes as shown in Fig. 3.7.

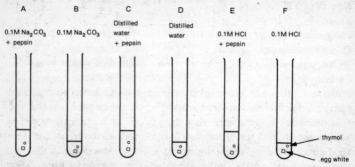

Fig. 3.7. *Apparatus for investigating the effect of pH on albumin digestion*

3. $2 \, cm^3$ of pepsin solution are added to A, C and E. Tubes B, D and F act as controls.
4. The tubes are kept at 35 °C for 24 hours and the cubes then examined.

Results

A and all the controls (B, D and F) show no signs of digestion. The cube is unchanged.

In C the cube has been slightly digested and the surface of the cube has a fuzzy appearance.

In E the cube has been totally digested and is no longer visible.

Conclusion

The enzyme pepsin requires acid conditions to digest egg albumin.

Here are two final points regarding enzymes:
1. Students sometimes write of enzymes being 'killed'. Enzymes cannot be killed, they are non-living. They can, however, be denatured.

2. Enzymes catalyse *all* the chemical reactions of metabolism, not just those concerned with digestion.

Definitions

Cell
Organelle
Tissue
Organ

Key Words

Inorganic molecules	Anabolism	Endoplasmic reticulum
Organic molecules	Nucleus	Ribosomes
Universal solvent	Cytoplasm	Cell wall
Carbohydrates	Cell (plasma) membrane	Vacuole
Proteins	Unicellular	Chloroplasts
Fats	Multicellular	Starch
Nucleic acids	Phagocytic	Centrioles
Movement	Autotrophic	Glycogen
Feeding	Photosynthesis	Monosaccharides
Nutrition	Heterotrophic	Disaccharides
Respiration	Locomotion	Polysaccharides
Excretion	Unlimited growth	Sugars
Growth	Nuclear membrane	Hydrolysis
Reproduction	Chromosome	Condensation
Sensitivity	Hereditary material	Amino acids
Metabolism	Chromatin	Fatty acids
Catabolism	Mitochondria	Glycerol

Exam Questions

1. (a) Make a labelled diagram to show the structure of a named unicellular animal. [5]
(b) How does the structure of this animal differ from:
(i) a simple animal cell such as from the lining of your cheek;
(ii) a named unicellular plant? [8]
(c) For each of the following organelles: mitochondrion, chloroplast, nucleus, plasma membrane, vacuole, give one function. [5]

[OXF]

2. *List three important differences between a bacterial cell and a named
protozoan.*   [6]

(See pages 307 and 326.)   [LON]

3. *What is meant by the terms*
(i)   carbohydrate
(ii)   fat
(iii) protein?   [6]
                                                                [O&C]

4. *(a) What are enzymes?*   [3]
(b) What are the main properties of enzymes?   [5]
*(c) Describe, with full practical details, an experiment you would perform
to find out how the rate of an enzyme-controlled reaction is related to
temperature.*   [10]
*(d) What results would you expect to obtain from your experiment and
how would you express them?*   [3]
                                                                [OXF]

5. $C_{12}H_{22}O_{11}$ *is the chemical formula for*
A. an amino acid
B. a protein
C. a fat
D. a carbohydrate
E. a nucleic acid.   [1]
                                                                [LON]

4. Movement of Substances Into and Out of Cells

Substances generally move into and out of cells by one of three processes: diffusion, osmosis or active transport. These processes are a popular topic with examiners and a clear understanding of them is essential.

As we have seen in Chapter 3, living organisms are largely made up of water. All living cells are surrounded by water and any substance that passes into or out of a cell must be in solution.

4.1. *Diffusion*

The molecules in a liquid or gas are constantly moving. Unless prevented from doing so by a barrier, they tend to move from a region where they are at a high concentration to a region where they are at a lower concentration. This movement continues until the molecules are evenly distributed. For example, if you drop a small amount of liquid dye into a container of water, the dye molecules will eventually spread throughout the container colouring the whole water. The dye molecules have spread from a region of high concentration to a region of low (initially zero) concentration. This movement of molecules is a physical process termed diffusion.

Definition: Diffusion *is the free movement of molecules of a substance from regions of high concentration to regions of lower concentration. The process continues until the molecules are evenly distributed.*

In an organism, many substances are in solution and diffusion is the usual way they move into, out of or between cells.

Examples of diffusion in living cells are:
1. Diffusion of oxygen and carbon dioxide (gas exchange) across the lining of alveoli in the lungs of mammals (see page 127).
2. Diffusion of oxygen and carbon dioxide (gas exchange) through the stomata of leaves in higher plants (see page 129).

In fact, wherever gas exchange occurs, diffusion of oxygen and carbon dioxide is occurring (for other examples see pages 117–22).

Any substance that can pass through a cell membrane will diffuse into or out of a cell if its concentration inside the cell is different to that outside.

As an illustration of this, let us look at a smooth muscle cell in a mammal. As the muscle cell respires it uses up oxygen and the oxygen concentration in the cell decreases. The concentration of oxygen outside the cell is higher so that oxygen will diffuse into the cell.

As the cell respires, it gives off carbon dioxide. The concentration of carbon dioxide in the cell will rise above that of its surroundings and carbon dioxide will diffuse out of the cell.

Glucose, another substance used up in cell respiration, will diffuse into the cell to replace glucose that has been broken down.

Think of any cell you like and try and work out what chemical processes are going on inside it. For example, in a mesophyll cell in the leaf of a green plant exposed to sunlight, photosynthesis as well as respiration is occurring (page 78). What substances are being used up by the cell and what substances are given off? By asking these questions you should be able to work out what substances are diffusing into or out of the cell and in what direction.

To return to our muscle cell, we can summarize the diffusion of substances into and out of the cell in a diagram:

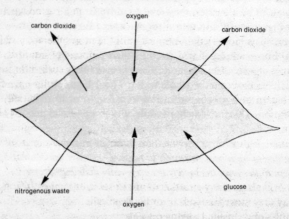

Remember, substances can only diffuse into or out of a cell if the cell membrane is permeable to them.

Finally, in dealing with diffusion, the term **concentration gradient** or **diffusion gradient** is often used. A concentration gradient is the difference in concentration between one region and another. Substances diffuse along a concentration gradient from a high to a low concentration. The greater (steeper) the concentration gradient the faster the rate of diffusion. Surfaces where gas exchange occurs are often designed to maintain a steep diffusion gradient so that diffusion occurs rapidly (see pages 117–27).

4.2. *Osmosis*

Osmosis is a special type of diffusion. It is the diffusion of water across a semi-permeable (sometimes known as a selectively permeable) membrane. Cell membranes are semi-permeable and it is by osmosis that water moves into, out of, and between cells. A semi- (selectively) permeable membrane is one that allows small molecules, e.g. gases and water, to pass across it, but not larger molecules, e.g. sugars and proteins.

Let us see how osmosis works. Look at Fig. 4.1 (over) which shows a simple apparatus used to demonstrate osmosis. Inside the thistle funnel is a strong solution of sugar while outside in the breaker is pure water. Stretched across the bottom of the funnel separating the sugar solution from the pure water is a semi-permeable membrane. This can be a natural membrane, e.g. a pig's bladder, or an artificial one, e.g. cellophane. Let us see what happens.

After a time the level of liquid in the thistle funnel will rise. Water has flowed by osmosis from the pure water, across the semi-permeable membrane and into the sugar solution. Why is this so?

What the water has done is simply diffuse along a concentration gradient from a high concentration of water (the pure water) to a low concentration of water (the sugar solution). Put another way, water is diffusing from a weak solution (of the sugar) to a strong solution.

The sugar molecules cannot diffuse from the sugar solution into the pure water because they are too large to pass through the membrane. Remember the membrane is semi-permeable, it will allow the water molecules through but not the sugar. Water molecules move through in both directions, but the water molecules move into the sugar solution faster than they move out. Thus there is an overall movement of water from the pure water into the sugar solution.

Definition: Osmosis *is the diffusion of water molecules through a semi-permeable membrane from a weak solution to a strong solution.*

You must be absolutely clear about this definition. Osmosis involves **water** diffusing through a **semi-permeable** membrane.

Let us return to the apparatus in Fig. 4.1. We could apply an external pressure to the sugar solution in the thistle funnel; a pressure that would be just sufficient to stop the flow of water into the sugar solution. This pressure is called the osmotic pressure of the solution. The stronger (more concentrated) the solution, the higher its osmotic pressure. Thus, water will diffuse from a solution of low osmotic pressure to one of higher osmotic pressure.

Strictly, the osmotic pressure of a solution is always measured relative to pure water and across a semi-permeable membrane.

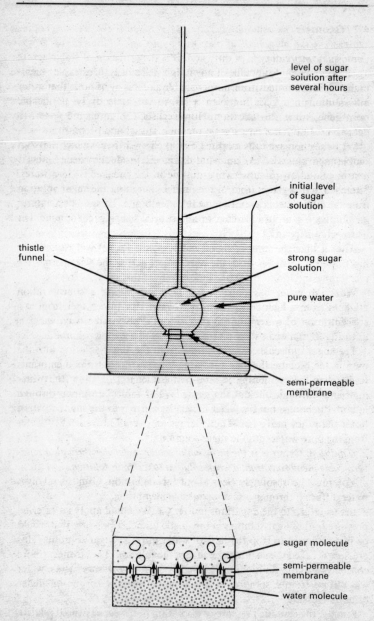

Fig. 4.1. *Demonstration of osmosis*

Definition: The osmotic pressure *of a solution is the pressure required to prevent the movement of pure water into the solution across a semi-permeable membrane.*

Really, we can only talk of a solution as having an osmotic pressure when it is separated from pure water by a semi-permeable membrane. A solution in a glass bottle or a cell in the body of an organism is obviously not in this situation. However, such solutions are potentially capable of exerting an osmotic pressure if separated from pure water by a semi-permeable membrane. For such solutions we say they have an **osmotic potential**, the potential to exert an osmotic pressure. Strictly, we should always use the term osmotic potential when dealing with body fluids or cell contents inside an organism. Throughout this book we will use the term osmotic potential (O.P.).

The ways that animal and plant cells respond to osmotic flows of water are very different. This is because plant cells are surrounded by a rigid cell wall whereas animal cells are not.

Osmosis and animal cells

The human red blood cell is a convenient animal cell in which to investigate osmosis. The osmotic potential of a red blood cell, as with other cells, is determined by the concentration of dissolved salts and dissolved organic substances it contains.

If a red blood cell is put in water, water will enter the cell by osmosis and the cell will swell up and may eventually burst.

If a red blood cell is put in a solution of greater concentration than itself, e.g. a strong salt solution, then the cell will lose water by osmosis and will shrink and wrinkle (Fig. 4.2).

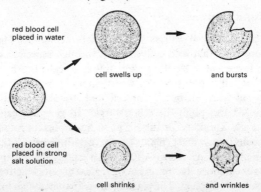

Fig. 4.2. *The effect on a red blood cell of changing the composition of the surrounding liquid*

Swelling up or shrinking can be damaging to a cell. To avoid this, animal cells are either surrounded by a liquid of the same osmotic potential as their own cell contents, or the cells have special mechanisms to regulate water content.

Multicellular animals maintain their body fluids at a relatively constant O.P. using a variety of methods: vertebrates do so using kidneys (pages 168–72). Single-celled animals, on the other hand, are able to actively pump water out using contractile vacuoles (page 167).

All these methods of regulating osmotic potential come under the term **osmoregulation** (page 167).

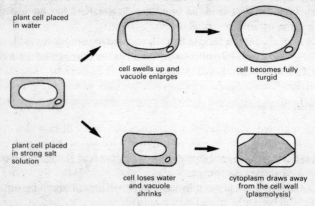

plant cell placed in water

cell swells up and vacuole enlarges

cell becomes fully turgid

plant cell placed in strong salt solution

cell loses water and vacuole shrinks

cytoplasm draws away from the cell wall (plasmolysis)

Fig. 4.3. *The effect on a plant cell of changing the composition of the surrounding liquid*

Osmosis and plant cells

A plant cell has a cell wall as well as a cell membrane. The cell wall is fully permeable (allows dissolved solids through as well as water) and it is rigid. It prevents the cell from bursting if the cell absorbs water.

If a plant cell is put in water, water enters by osmosis and the cell swells up but does not burst. The cell wall stretches slightly but does not break. At this point the cell is turgid. In general, plant cells are slightly turgid. Turgidity, or turgor, plays an important role in support in herbaceous (non-woody) plants (see page 320). When all the cells in a particular plant tissue absorb water, each cell swells and presses out against its neighbours. The result is a rigid mass of cells which can support the soft parts of the plant. The packing tissue, parenchyma, in particular plays this role.

If a plant cell is put in a strong sugar solution water passes out of the cell by osmosis and the cell loses its turgor and becomes flaccid. Under experimental conditions, if the external solution is strong enough, the cytoplasm shrinks and eventually pulls away from the cell wall, a process called plasmolysis. Plasmolysis does not normally occur in a living plant. A loss of turgidity, however, may occur, particularly in dry weather and drought conditions, and accounts for the wilting (drooping) of plant tissues.

Osmosis plays a key role in three important processes in plants:
1. Water is taken up into root hairs from the soil and then travels to the root xylem by osmosis (page 152).
2. Water travels from the xylem into the mesophyll cells of the leaf by osmosis (page 152).
3. Stomata (page 65) open and close by a change in turgidity of their guard cells. Increased turgidity is achieved by increasing the osmotic potential of the cell contents so that water flows in by osmosis. When this happens the stomatal pore opens. When the O.P. of the guard cells is lowered, water is drawn out, the cells become flaccid, and the stomatal pore closes.

4.3. *Active Transport*

Active transport is a chemical process which results in a movement of particles in an opposite direction to that expected by diffusion. It occurs when molecules are taken across a membrane from a region of low concentration to a region of higher concentration, i.e. against a concentration gradient. Protein 'carrier' molecules within the cell membrane are thought to carry the substances across.

Active transport, as its name implies, is an active process and requires energy. This energy is obtained from respiration and any factors which locally influence the rate of respiration also influence the rate of active transport. Thus, low temperature, lack of oxygen and respiratory inhibitors such as cyanide, all inhibit active transport.

Three well-known examples of active transport are:
1. The active uptake of certain mineral salts by the root hair cells of a plant (see page 159).
2. The selective reabsorption of substances by mammalian kidney tubules (see page 171).
3. The absorption of digested food from the mammalian ileum is aided by active transport (see page 90). Active transport, in this case, is used to aid diffusion, by speeding up the rate at which substances are absorbed.

Read back over the chapter once again and then answer the questions below.

Definitions

Diffusion
Osmosis
Osmotic pressure

Key Words

Concentration (diffusion) gradient
Semi- (selectively) permeable membrane
Osmotic potential
Turgidity (turgor)
Plasmolysis
Active transport

Exam Questions

1. (a) *What do you understand by the term osmosis?*   [5]
(b) *How does diffusion differ from this?*   [3]
[O&C]

2. *In a single-celled organism living in water, explain how diffusion is involved in:*
(i) *movement of oxygen*   [2]
(ii) *excretion of dissolved substances.*   [2]
[SUJB]

   (Look at page 117 if you need help in answering this question.)

3. (a) *Describe an experiment to demonstrate osmosis.*
(b) *What part does this process play in the uptake of water by a root-hair cell?*
(c) *Why are contractile vacuoles generally present in freshwater protozoa but absent from marine protozoa?*   [25]
[LON]

   (Look at pages 152 and 167 if you need help in answering parts (b) and (c) of this question.)

5. Feeding (Nutrition)

Nutrition is the study of food and feeding processes. Food is the material from which organisms obtain:
1. the energy
2. the raw materials
to construct, maintain and repair the body.

5.1. Types of Nutrition

There are two main types of nutrition: **autotrophic** (holophytic) nutrition, typical of green plants, and **heterotrophic** nutrition, typical of animals and non-green plants.

Autotrophic organisms make their own organic food from simple inorganic substances, using the energy from sunlight – a process called **photosynthesis**.

Heterotrophic organisms get their organic food 'ready-made'. Directly or indirectly, they are completely dependent on plants for it. All their organic food has originally been made by a green plant.

Heterotrophic nutrition can be divided into three types:

1. **Holozoic nutrition**: this is typical of most animals, ranging from *Amoeba* (page 326) to man. It involves five stages:
(a) Complex organic food is taken into a cavity within the body (**ingestion**).
(b) Within the cavity, organic food is broken down into simple soluble substances (**digestion**).
(c) The soluble products of digestion are taken up into the organism (**absorption**).
(d) These soluble substances are incorporated into the protoplasm of cells (**assimilation**).
(e) Indigestible waste is removed from the body (**egestion**).
2. **Saprophytic nutrition**: this type of nutrition is shown by many non-green plants, including certain fungi, e.g. *Mucor* (page 314) and certain bacteria (page 307). The saprophytic organism (saprophyte) secretes digestive enzymes on to dead or decaying organic matter. The organic food is digested and the soluble products of digestion are absorbed.

Saprophytic nutrition differs from holozoic nutrition in that digestion occurs *outside* the organism and the organic food is always obtained from dead or decaying material.

Saprophytes play a very important role in the cycling of materials in nature (see page 290).

3. **Parasitic nutrition**: a parasitic organism (parasite) lives in or on another live organism (its **host**) from which it obtains its organic food. Examples of parasites are viruses (page 305), certain bacteria (page 307), the fungus *Phytophthora* (page 315) and tapeworms (page 330).

5.2. *Photosynthesis*

The process by which green plants make organic compounds from simple inorganic compounds is called **photosynthesis**. It can be represented by the following equation:

$$\text{carbon dioxide} + \text{water} \xrightarrow[\text{chlorophyll}]{\text{sunlight}} \text{glucose} + \text{oxygen}$$

$$6CO_2 + 6H_2O \rightarrow C_6H_{12}O_6 + 6O_2$$

This equation is really an oversimplification:

1. Glucose is not the only product of photosynthesis, though it is the main one. Other sugars are formed which can then be converted to fats or, by combining with mineral salts, form amino acids and vitamins. Photosynthesis is the source of all organic compounds in a plant.

2. Photosynthesis is not a one-step chemical reaction, but is really a series of over forty reactions, each reaction having a specific enzyme. All green parts of a healthy plant can carry out photosynthesis. The green coloration is due to the presence of the pigment **chlorophyll**, which traps sunlight energy. The chlorophyll converts this energy to chemical energy, which is used to bind atoms and small molecules together to make organic molecules.

In flowering plants (page 319) the leaves are the main sites of photosynthesis. Photosynthesis also occurs in the green stems of herbaceous plants (those flowering plants without woody stems, page 320).

The leaf of a herbaceous dicot (page 319) is made up of a **midrib** and **lamina** attached to the stem by a stalk or **petiole**. A vein (vascular bundle) runs through the midrib and branches off into smaller veins in the lamina (see Fig. 5.1).

Within a cell of the leaf, **chloroplasts** are the actual sites of photosynthesis. It is these organelles (page 39) which contain chlorophyll together with the enzymes which catalyse the reactions of photosynthesis.

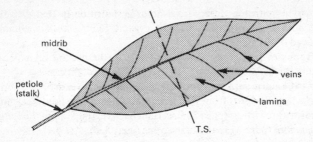

Fig. 5.1. *A leaf from a dicot (see page 319)*

In transverse section (sometimes called a vertical section), the lamina of the leaf is seen to have four tissue layers covered on top and bottom by a thin waxy cuticle (see Fig. 5.2.). Cells of the *palisade* layer contain the largest number of chloroplasts and it is here that the bulk of photosynthesis occurs. The **spongy mesophyll** cells contain fewer chloroplasts and the cells are separated by large air spaces which allow diffusion of gases through the leaf. The single layer of cells (**epidermis**) at the top and bottom of the lamina contain no chloroplasts except in the **guard cells** of the **stomata** (singular: stoma). Stomata are pores (holes) in the leaf surface which can be opened or closed by changes in turgidity of the guard cells which border them (see page 61). The pores allow carbon dioxide and oxygen to be exchanged between the leaf and the atmosphere. Stomata are usually found only on the lower surface of the leaf, away from the sun's direct rays. This reduces water loss by evaporation and diffusion through the stomatal pores.

Questions on leaf structure, and how leaf structure is related to photosynthesis, are popular with examiners. For example:

Q. Make a labelled diagram to show the arrangement of structures in the leaf of a flowering plant as seen in vertical section. Indicate on your diagram the distribution of chloroplasts. [9, 3]
[OXF]

Fig. 5.2 provides an effective answer.

Q. In what ways are the leaves of plants adapted for photosynthesis? [6]
[O&C]

For photosynthesis to occur, carbon dioxide, water and light must come together at the site of photosynthesis. Let us examine how each of these factors reaches the chloroplasts in the leaf cells:

1. **Carbon dioxide** diffuses from the atmosphere through the stomata and into the air spaces of the spongy layer. The gas dissolves in the

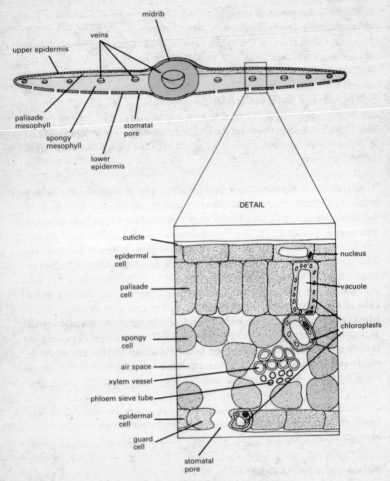

Fig. 5.2. *Transverse (vertical) section through the leaf of a dicot*

moisture on the surface of mesophyll cells and diffuses to the chloroplasts. The following features of the leaf encourage this inward diffusion of carbon dioxide:

(a) The leaf is flat and has a large surface area.

(b) Stomata allow carbon dioxide to enter the leaf from the atmosphere.

(c) Air spaces in the spongy mesophyll allow carbon dioxide to diffuse to the palisade cells.

(d) The leaf is thin so diffusion distances are short.

2. **Water** is delivered to the leaf in the xylem vessels of the veins. The water originally enters the plant by osmosis through root hairs (page 152). The water then rises up the xylem vessels of the stem by a combination of root pressure, capillarity and transpiration (page 153). As cells in the leaf use up water in photosynthesis, more water is drawn from the xylem vessels by osmosis (page 152). The following features enable delivery and retention of water in the leaf:

(a) Xylem vessels in the veins deliver water to all parts of the leaf.

(b) The waxy cuticle covering the leaf reduces water loss by evaporation.

(c) The stomata, which allow gas exchange to occur, are found in larger numbers on the lower surface, away from direct sunlight. This reduces water loss by evaporation.

3. **Light**: the following features maximize the available light which reaches the chloroplasts in the palisade layer:

(a) The palisade cells are close to the upper surface of the leaf, directed towards the sunlight.

(b) The palisade cells are elongated towards the direction of sunlight. This reduces the number of cross walls light has to pass through before reaching chloroplasts.

(c) The large surface area of the lamina exposes a very large area of chlorophyll-containing cells to the light.

(d) The leaves are held at right angles to the light so that the upper surface receives maximum light.

(e) The leaves are arranged on the stem so that one leaf does not over-shadow another. This arrangement is called a **leaf mosaic**.

The above points should enable you to answer the following question which leads on from the diagram question answered in Fig. 5.2:

Q. Explain fully how the raw materials required for photosynthesis reach the chloroplasts. [13]

[OXF]

NOTE. Light is not a raw material.

5.3. *The Fate of the Products of Photosynthesis*

Glucose and other sugars

1. Much of the glucose is converted to starch for temporary storage in the leaf. At night the starch is broken down to sucrose and transported round the plant in the phloem (page 149).

2. In the leaf and throughout the plant, sugars are broken down in respiration to release energy.

3. In growing regions, glucose is condensed to cellulose to make cell walls.

4. In the leaf, some sugars are combined with nitrates to form amino acids which are later incorporated in proteins.

5. In the leaf, some sugars are used to make fats and vitamins.

Remember, all cells in the plant will need a supply of organic substances. Photosynthesis is the original source of these. The substances are transported in the phloem to cells throughout the plant (see page 160).

Oxygen

1. Used in respiration, or
2. Excreted through the stomata.

5.4. *Photosynthesis Experiments*

In a healthy plant, an adequate supply of the following is required for photosynthesis to take place:

    carbon dioxide
    water
    light
    chlorophyll
    plus, a suitable temperature.

These requirements can be demonstrated in a series of experiments. In such experiments, the presence of starch in a leaf is normally taken to indicate that photosynthesis has occurred. The assumption is that starch is a product of photosynthesis. This is generally the case: glucose formed in photosynthesis is converted to starch for temporary storage in the leaf.

Testing for starch

A green leaf is removed from a plant and the following steps are taken:
1. The leaf is placed in boiling water for five minutes to fix (kill) the leaf tissue and break down the waxy cuticle.
2. The leaf is transferred to boiling ethanol (alcohol) heated in a water bath to reduce the danger of fire. The ethanol removes chlorophyll. This

decolorizes the leaf so that colour changes used in the starch test are not masked by the green colour of chlorophyll.

3. The decolorized leaf is put in hot water to soften it.
4. The softened leaf is spread out on a white tile.
5. It is covered for 1 minute with dilute iodine solution.
6. The iodine solution is then rinsed off with water.

A blue-black coloration indicates the presence of starch. If the brown colour of iodine persists, the test is negative for starch.

In experiments where the presence of starch is used to indicate photosynthesis, the leaves of the plant should be destarched before the experiment is begun. This is done by keeping the plant in darkness for 48 hours.

An experiment which demonstrates the need for light in photosynthesis

Method
1. Use a destarched plant.
2. Set up control and test leaves as shown in Fig. 5.3. For the test, the leaf is sandwiched between two sheets of dark paper, so that both top and bottom surfaces of the leaf are covered. For the control, transparent or translucent paper is used.

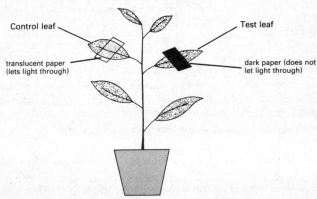

Fig. 5.3. *Apparatus testing the need for light in photosynthesis*

3. Give the plant water and leave in daylight or artificial light for six hours.
4. Remove the test and control leaves and test them for starch (see starch test above).

Results
1. The covered strip on the control leaf turns blue-black with iodine (positive for the starch test).
2. The covered strip on the test leaf does not turn blue-black (negative for the starch test).

Conclusion
The absence of starch in the covered strip of the test leaf indicates that in the absence of light photosynthesis does not take place.

Light is needed for photosynthesis to occur.

An experiment which demonstrates the need for carbon dioxide in photosynthesis

Method
1. Use a destarched plant.
2. Set up the control and test leaves as shown in Fig. 5.4.

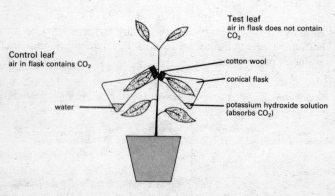

Fig. 5.4. *Apparatus testing the need for CO_2 in photosynthesis*

3. Give the plant water and leave in daylight or artificial light for six hours.
4. Remove the test and control leaves and test them for starch (see starch test above).

Results
1. The control leaf turns blue-black with iodine (positive for the starch test).
2. The test leaf does not turn blue-black (negative for the starch test).

Conclusion

The absence of starch in the test leaf indicates that in the absence of carbon dioxide photosynthesis does not take place.

Carbon dioxide is needed for photosynthesis to occur.

An experiment which demonstrates the need for water in photosynthesis

Such an experiment cannot easily be performed. The need for water in photosynthesis cannot be tested by depriving the leaf or plant of water. Without water, living tissue will die. In practice, sophisticated labelling experiments are required to show the need for water in photosynthesis (see section 5.5).

An experiment which demonstrates the need for chlorophyll in photosynthesis

A variegated plant, such as a variegated strain of geranium, is used. A variegated plant has white or yellow chlorophyll-free patches on its leaves.

Method

1. Use a destarched plant.
2. Give the plant water and leave in daylight or artificial light for six hours.
3. Remove a leaf and make an accurate drawing of its colour patterning.
4. Test the leaf for starch.

NOTE. The green chlorophyll-containing part of the leaf is the control and the yellow or white chlorophyll-free part is the test.

Results

When the distribution of chlorophyll is compared with the results of the starch test (see Fig. 5.5), typical findings are:

1. Those patches originally green give a blue-black coloration with iodine (positive for the starch test).
2. Those patches originally white or yellow do not turn blue-black with iodine (negative for the starch test).

Conclusion

The absence of starch in the chlorophyll-free part of the leaf indicates that in the absence of chlorophyll, photosynthesis does not take place.

Chlorophyll is needed for photosynthesis to occur.

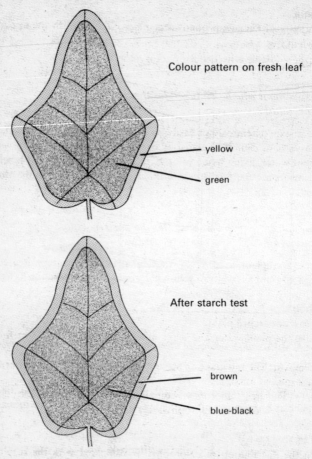

Fig. 5.5. *Variegated ivy leaf before and after testing for starch*

An experiment which shows that a suitable temperature is required for photosynthesis to take place

In this experiment, rate of photosynthesis is measured. In theory, changes in the amounts of either the raw materials or products of photosynthesis could be used to measure the rate of reaction. In practice, it is easiest to measure the volume of gas (oxygen) given off by the plant in a given length of time. Oxygen is most easily collected over water, and so an aquatic plant is used.

Method

1. A sprig of Canadian pondweed, *Elodea*, is placed in the apparatus shown in Fig. 5.6.

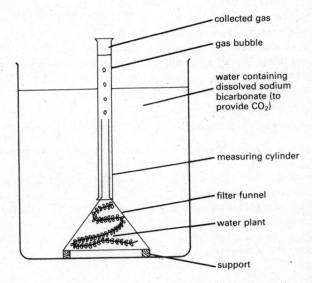

Fig. 5.6. *Apparatus for testing the effect of temperature on rate of photosynthesis*

2. The plant is exposed to a standard light source for a set time and the bubbles of gas given off during this time are collected in a measuring cylinder. The total volume of collected gas is recorded.

3. The experiment is conducted at different temperatures, starting at 15 °C and repeated at 10 °C intervals up to 45 °C. The apparatus is kept in a water bath to minimize temperature fluctuations.

Results

The rate of photosynthesis (as measured by volume of gas given off per hour or minute) increases with rise in temperature, up to 35 °C. At 45 °C very little gas is collected.

Conclusions

Rate of photosynthesis, as measured by gas evolved, is influenced by temperature. Between 15 °C and 35 °C, a rise in temperature increases the rate of photosynthesis. This is typical of most chemical reactions. When the temperature is raised, the reaction is speeded up. At 45 °C, photosynthesis is inhibited because the enzymes involved in photosynthesis are being denatured (see page 49).

NOTE. There are certain complications with this experiment:
1. Solubility of gases decreases with rise in temperature.
2. Gases increase in volume with rise in temperature.
3. Not all gas collected is necessarily oxygen.
4. Some of the oxygen given off in photosynthesis will be used up in respiration.

An experiment which shows that oxygen is given off in photosynthesis

The apparatus from the temperature experiment above can be used. In the presence of light, the bubbles of gas given off by a pondweed can be shown to contain largely oxygen. The collected gas will relight a glowing splint, a test for oxygen.

If the experiment is repeated in the absence of light, the collected gas will not relight a glowing splint.

The two results taken together indicate that oxygen is given off by a photosynthesizing plant.

5.5. *The Mechanism of Photosynthesis*

NOTE. This is not required by all examining boards. Check the Contents list for your exam board requirements.

As mentioned earlier, the simple equation for photosynthesis gives no indication of the complexity of the reactions involved. Each reaction requires a specific enzyme, and all these enzymes are contained within chloroplasts. Photosynthesis occurs in two main stages, the light and dark reaction.

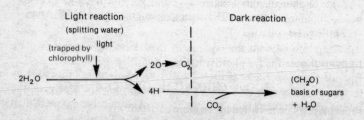

1. **Light reaction**: during this stage chlorophyll traps light energy and converts it to chemical energy. This energy is used to split water into oxygen and hydrogen. Oxygen is released as a waste product, and the hydrogen is used in the second phase of photosynthesis, the dark reaction.
2. **Dark reaction**: this stage does not require light, but depends on a supply of hydrogen from the light reaction. Hydrogen combines with carbon dioxide to form sugars, which may afterwards be converted to other types of organic molecule.

Labelling experiments

The principle of such experiments lies in labelling one of the raw materials in a reaction and then examining the products of the reaction to see which one contains the label. Details of the reaction pathway and mechanism can be learnt in this way.

Experiments using heavy oxygen (^{18}O)

These experiments provide direct evidence that the oxygen gas released during photosynthesis comes from water and not carbon dioxide. The experiments also demonstrate that water is a reactant in the photosynthesis reaction and is therefore necessary for photosynthesis to occur (see page 68).

When water labelled with ^{18}O is supplied to a photosynthesizing plant, the label turns up in the oxygen released by the plant. When carbon dioxide labelled with ^{18}O is supplied, the label appears in the sugars and starch formed during photosynthesis. These experiments indicate that in photosynthesis, the oxygen in water is released as molecular oxygen, while the oxygen in carbon dioxide is incorporated into sugars.

carbon dioxide +   water   → glucose +   oxygen
    (labelled with ^{18}O)   (labelled with ^{18}O)

carbon dioxide + water→ glucose   + oxygen
(labelled with ^{18}O) (labelled with ^{18}O)

Taking into account the findings of these labelling experiments, the traditional equation for photosynthesis:

$$6CO_2 + 6H_2O \rightarrow C_6H_{12}O_6 + 6O_2$$

should strictly be rewritten in the more correct form:

$$6CO_2 + 12H_2O \rightarrow C_6H_{12}O_6 + 6O_2 + 6H_2O$$

Why? Because if all the oxygen given off comes from water, then the traditional equation is unbalanced. Six oxygen atoms in water are giving rise to twelve oxygen atoms in the oxygen gas. The rewritten form balances the equation.

Experiments using radioactive carbon (^{14}C)

The role of carbon dioxide in photosynthesis has been established using the ^{14}C label. If carbon dioxide labelled with ^{14}C is supplied to a photosynthesizing plant, the label is first found in sugars and later in a range of organic compounds. This type of experiment shows that the carbon in carbon dioxide is incorporated into carbohydrates, fats and proteins.

5.6. *Limiting Factors and Photosynthesis*

In section 5.5 we examined the conditions which are necessary for photosynthesis to occur. If any one of these factors is in short supply, the rate of photosynthesis may be slowed down. The factor in shortest supply, and therefore the one controlling the rate of the reaction, is called the **limiting factor**.

The principle of limiting factors can be illustrated by placing a plant in increasing light intensities and measuring its rate of photosynthesis, whilst keeping the temperature and carbon dioxide concentration constant.

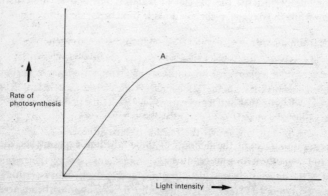

Fig. 5.7. *The effect of light intensity on rate of photosynthesis*

The results plotted in Fig. 5.7 indicate that up to point A an increase in light intensity produces an increase in the rate of photosynthesis, but after that point, the rate levels off. This levelling-off could have a number of causes:

1. The temperature may not allow the reaction to proceed any faster.
2. There may be an inadequate supply of carbon dioxide.
3. The products of the reaction may be building up and preventing any further action.
4. This may be the maximum rate at which the 'photosynthetic machinery' can work under any conditions.

To find out whether a particular factor is limiting the reaction, you increase the supply of that factor and see if the rate increases. For example, if the experiment above is repeated at a higher concentration of carbon dioxide and the rate of photosynthesis increases (see Fig. 5.8), then carbon dioxide was originally acting as a limiting factor.

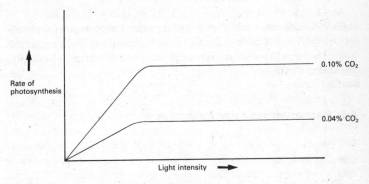

Fig. 5.8. *The effect of light intensity on rate of photosynthesis at different CO_2 concentrations*

If we consider a plant growing outdoors in a temperate climate, carbon dioxide concentration in the atmosphere varies little (a constant 0·04 per cent), whereas light and temperature vary with time of day and season of the year.

Curve A of Fig. 5.9 shows the rate of photosynthesis for a plant at different times on a hot, sunny day in summer. For some time after dawn, light and temperature are probably the limiting factors; during the middle hours of the day carbon dioxide is limiting; and at dusk, light is limiting.

On a cool sunny day (curve B) temperature may be limiting throughout the day, and light intensity limiting at dawn and dusk.

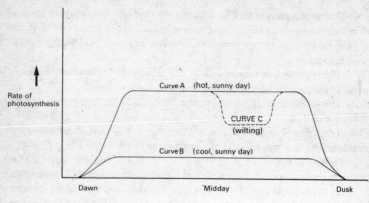

Fig. 5.9. *Rates of photosynthesis at different times of day, under different conditions*

Any factor which causes stomata to close, thus stopping the flow of carbon dioxide into the leaf, will limit the rate of photosynthesis, regardless of the level of carbon dioxide in the atmosphere. Wilting, due to lack of water, will reduce the rate of photosynthesis (curve C) by causing the stomata to close.

5.7. *Photosynthesis and Respiration*

1. Photosynthesis is essentially the reverse of respiration. In photosynthesis, organic substances are built up and provide a store of chemical energy. In respiration, organic substances are broken down to release energy.

2. Photosynthesis uses the products of respiration while respiration uses the products of photosynthesis:

$$\text{carbon dioxide} + \text{water} + \text{energy} \underset{\text{respiration}}{\overset{\text{photosynthesis}}{\rightleftharpoons}} \text{glucose} + \text{oxygen}$$

3. Green plants (like all living things) respire all the time (both in the light and in the dark).

4. During daylight, plants photosynthesize as well as respire. For most of the daytime, photosynthesis proceeds at a much faster rate than respiration (see Fig. 5.10). More than enough food is made during the day to last through the night.

We can apply the above points to the events which occur in a leaf

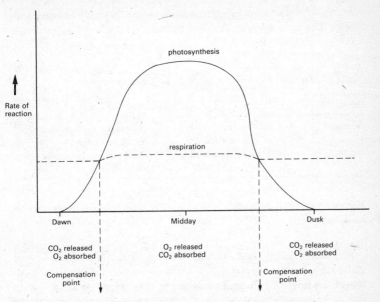

Fig. 5.10. *Rates of respiration and photosynthesis over a 24-hour period*

over a 24-hour period. In darkness, respiration alone occurs and the leaf takes in oxygen and releases carbon dioxide through the stomata. As dawn breaks, photosynthesis begins and the release of carbon dioxide slows down and stops as carbon dioxide is used up in photosynthesis. At the same time, absorption of oxygen from the air slows down and stops as oxygen is produced in photosynthesis. At dawn and dusk each day there is a **compensation point**, where the reactions of photosynthesis and respiration occur at the same rate (see Fig. 5.10). At the compensation point there is no overall movement of oxygen or carbon dioxide into or out of the leaf. During the middle part of the day, photosynthesis exceeds respiration and the leaf takes up carbon dioxide and gives off oxygen.

5.8. *The Mineral Salt Requirements of Plants*

In addition to carbon dioxide and water needed for photosynthesis, plants require a supply of mineral salts. These, together with sugars made in photosynthesis, are used to manufacture proteins and vitamins.

Mineral salts enter the root hairs of the plant as ions (see page 159).

They travel through the xylem to the leaf, where they are combined with sugars. The main elements supplied by mineral salts, together with their roles, are shown in Table 5.1

Table 5.1. *Elements required by plants and supplied in mineral salts*

Element		
Nitrogen (N)	As nitrate (NO_3^-), e.g. in KNO_3	In proteins and nucleic acids (DNA and RNA)
Sulphur (S)	As sulphate ($SO_4^=$), e.g. in $MgSO_4$	In proteins and nucleic acids (DNA and RNA)
Phosphorus (P)	As phosphate ($PO_4^=$), e.g. in K_2PO_4	In enzymes, ATP and nucleic acids
Calcium (Ca)	As calcium ion (Ca^{++}), e.g. in $CaCO_3$	'Gum' (middle lamella) between adjacent cell walls
Iron (Fe)	As ferrous ion (Fe^{++}), e.g. in $FeSO_4$	In enzymes involved in making chlorophyll
Magnesium (Mg)	As magnesium ion (Mg^{++}), e.g. in $MgSO_4$	Part of chlorophyll molecule
Potassium (K)	As potassium ion (K^+), e.g. in K_2PO_4	Activates enzymes involved in respiration and photosynthesis

In addition, minute quantities of so-called 'trace' elements such as zinc (Zn), copper (Cu) and manganese (Mn) are needed for healthy growth. Large quantities are often poisonous.

Water culture experiments

The effect of a deficiency (lack) of a particular element on growth can be shown by 'water culture' experiments in which plants are grown in aerated solutions, rather than in soil.

For example, a batch of barley seedlings are grown in a series of solutions in glass bottles. A control plant is supplied with the complete range of mineral salts. A series of test plants are grown in solutions in which one of the key elements is missing. In all other respects, the control and test plants are grown under the same set of conditions. After a few weeks the seedlings are examined and the growth of seedlings in test solutions compared with that of the control.

Typical results are as follows:

Experimental conditions	Appearance of plant		
	Stem	Leaves	Roots
Control (seedling grown in a solution containing N, S, P, Ca, Fe, Mg, K)	Tall, sturdy	Dark green	Normal growth
As control but no N	Thin and weak	Pale green or yellow	Normal
As control but no S	Thin and weak	Pale green or yellow	Normal
As control but no P	Short	Small and purple	Poor growth
As control but no Ca	No growth	–	Poor growth
As control but no Fe	Thin and weak	White	Normal
As control but no Mg	Normal	Yellow	Normal
As control but no K	Normal	Yellow with dead spots	Normal

In setting up water culture experiments certain precautions are necessary:

1. The water used to make up the stock solutions should, as far as possible, be pure.
2. The bottles and solutions should be sterilized to reduce the likelihood of fungal or bacterial growth.
3. The bottles should be wrapped in dark paper to prevent algal growth.

5.9. *Holozoic Nutrition*

All animals feed on ready-made organic food supplied by green plants. Animals cannot photosynthesize and are unable to make their own organic materials from simple inorganic molecules.

Carnivores (e.g. the dog, page 99) feed mainly on animals.

Herbivores (e.g. the sheep, page 100) feed on plants.

Omnivores (e.g. man, page 99) feed on plants and animals.

Before food can properly enter the body it must pass across a cell membrane. In the case of animals, complex organic food can enter cells of the body only if first broken down into small, soluble molecules. The stages involved in this process are as follows:

1. **Ingestion**: the intake of complex organic foods into a cavity within the body.

2. **Digestion**: the breakdown of complex organic foods into small, soluble molecules.

3. **Absorption**: the uptake of soluble food substances into the body across a cell membrane.

4. **Assimilation**: the uptake and use of soluble food substances by cells in the body.

5. **Egestion**: the removal of undigested food from the body. Do not confuse egestion with excretion (page 165) or secretion (below).

6. **Secretion**: the production of useful chemical substances by a cell or gland.

In mammals, the alimentary canal (gut) is responsible for the ingestion, digestion, absorption and egestion of food. The gut is modified according to the diet of the animal (see sections 5.11 and 5.15).

The breakdown of foods in the gut involves both mechanical and chemical action. The teeth and tongue physically break down food by chewing (mastication). Digestive enzymes chemically break down foods by hydrolysis (see page 44). These enzymes are found in juices secreted by glands in the gut wall and by the pancreas.

Questions on digestive enzymes, together with conditions under which they operate, are very popular. Turn back to page 51 and then try the following question, using a digestive enzyme as your example.

Q. (a) What are the effects of a change in
(i)   temperature
(ii) pH, upon the rate of action of any named *enzyme*   [7]
(b) Describe with full experimental details how you would investigate either of these effects.
                                                                                      [12]
                                                                                   [OXF]

We will now look at the human diet and alimentary canal. Remember, man is an omnivore.

5.10. *The Human Diet*

Humans require a **balanced diet**. This is one that supplies the different types of food in adequate amounts and correct proportions, and provides the body with sufficient energy for its needs. A balanced diet maintains healthy active life and, where necessary, growth.

The components of a balanced diet are:

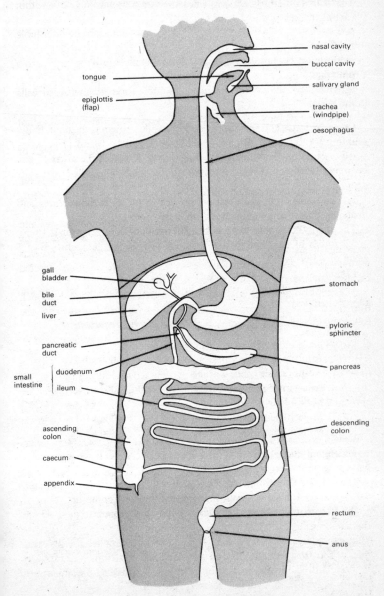

Fig. 5.11. *The human alimentary canal (gut) and associated structures*

a sufficient quantity of energy (measured in kilojoules)
carbohydrates
fats
proteins
vitamins
minerals
water
roughage.

Carbohydrates and fats supply energy, while protein is the main body-building food. Of the 20 amino acids which can be found in proteins, 8 are essential in the human diet. From these 8 any of the other 12 can be made. Proteins rich in essential amino acids are called first-class proteins.

Fat contains twice as much energy per gram as carbohydrates and proteins. This is why it is such a good store of energy.

The energy content of foods is measured in kilojoules per gram (kJ/gm).

4·2 joules of heat energy are required to raise the temperature of 1 gram of water by 1 °C.

1,000 joules = 1 kilojoule.

The amount of energy required in the diet varies with sex, age and physical activity.

	Energy requirement (*average*)
Adult male office worker	12,000 kJ/day
Adult female office worker	9,500 kJ/day
Adult male manual worker	15,000–20,000 kJ/day
Male 15 years old	11,000 kJ/day
Male 10 years old	9,000 kJ/day
Male 5 years old	6,500 kJ/day

Too much energy-giving food will cause overweight, while too little will cause underweight.

The detailed nutritional requirements for a human are shown in Tables 5.2, 5.3 and 5.4. Short-answer questions on these are common.

Questions on food tests are also popular. Refer back to page 47 and then answer this question.

Q. Describe, for each of the following, one *test you could carry out in order to demonstrate their presence in food materials:*

(i)   *reducing sugar, e.g. glucose*

(ii) *fats*

(iii) *proteins.*

[8]

[o&c]

Table 5.2. *The different parts of the human diet*

Type of food	Good sources	Main uses in the body
Carbohydrates:		
Starch	Potatoes, flour products	As an energy source
Disaccharides	Table sugar, milk	Stored as glycogen
Monosaccharides	Honey, fruit	
Proteins:		
First-class proteins (rich in essential amino acids)	Meat, eggs, milk, cheese	Structural material in cells. Enzymes. General growth and repair of tissues
Second-class proteins	Peas, beans	
Fats (lipids)	Unskimmed dairy products; e.g. butter. Plant oils	As an energy source. As structural material in cell membranes. Stored as fat in adipose tissue

Vitamins

At least twenty vitamins are required in minute quantities. Some are listed in Table 5.3.

Minerals

About fifteen different minerals are required in very small quantities. The major ones are listed in Table 5.4.

Water	Liquids and most foods	Body 60–80 per cent water. All chemical reactions take place in it
Roughage	Cereals, vegetables	Indigestible. No food value. Stimulates muscular movements (peristalsis) of gut wall. Lack of roughage can lead to constipation

Table 5.3. *Vitamins in the human diet*

Vitamin	Good sources	Main uses in the body
A (fat soluble)	Butter, milk, certain vegetables, e.g. carrots	Maintaining health of mucous membranes. Lack of vitamin (deficiency) increases likelihood of disease infection. Forms part of visual pigment in rod cells of eye (page 194). Lack of vitamin leads to night blindness
B (water soluble) There are over ten B vitamins, two of which are thiamine (B_1) and niacin (B_2)	Whole grain of cereal, such as in wholemeal bread. Liver, yeast	Thiamine and niacin play a role in energy release (respiration). Lack of thiamine leads to paralysis (**beri-beri**). Lack of niacin leads to gut disorders and the skin disease **pellagra**
C (water soluble) Destroyed by boiling or prolonged exposure to sunlight	Fresh citrus fruits. Certain fresh vegetables, e.g. potatoes. Milk	Repair of tissue damage. Maintaining health of skin and blood vessels. Lack of vitamin leads to **scurvy** (bleeding gums, poor wound healing, and bleeding under skin)
D (fat soluble)	Liver, fish-liver oil, dairy products. Also manufactured by the skin when exposed to u.v. light	Used in bone formation. Lack of vitamin leads to skeletal deformity such as **rickets** in children

Table 5.4. *Minerals in the human diet*

Mineral	Good sources	Main uses in the body
Calcium (Ca)	Dairy products, bread, fruit and vegetables	Bone and teeth formation. Lack of mineral leads to brittle bones and teeth
Iodine (I)	Table salt. Sea foods	In thyroid gland hormone, thyroxine. Lack of mineral leads to thyroid gland swelling (goitre)
Iron (Fe)	Liver, egg yolk	Formation of haemoglobin (see page 126)
Phosphorus (P)	Dairy products	Bone and teeth formation. Lack of mineral leads to brittle bones and teeth
Potassium (K) Sodium (Na)	Plant material Table salt	Important constituents of blood and cells. A proper balance of these is required for proper functioning of nerves and muscles

5.11. *The Human Alimentary Canal (Gut)*

The alimentary canal (gut) is a muscular tube running from mouth to anus (see Fig. 5.11). Different parts of the tube serve different functions. From mouth to duodenum, the gut is mainly concerned with the physical breakdown and chemical digestion of food. Absorption occurs in the ileum and colon. The last section of the gut, the rectum, prepares indigestible waste for disposal (egestion).

The gut wall has two layers of muscle, circular and longitudinal. These muscle layers act against each other to produce waves of contraction called **peristalsis** (see Fig. 5.12). By peristalsis, food is pushed along through the gut.

Mucus is secreted by the gut lining throughout its length and lubricates and protects the lining against abrasion.

The functions of the different parts of the gut can be summarized as follows:

Fig. 5.12. *Peristalsis in the oesophagus*

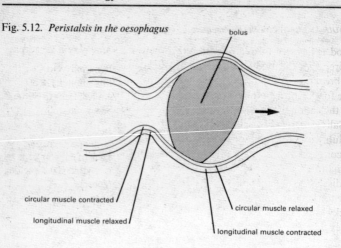

circular muscle contracted

longitudinal muscle relaxed

circular muscle relaxed

longitudinal muscle contracted

Table 5.5. *The functions of the different regions of the human alimentary canal (gut)*

Region	Functions
Mouth (buccal cavity)	Ingestion, mastication (chewing), bolus (food ball) formation, some starch digestion
Oesophagus	Delivery of food from mouth to stomach
Stomach	Storage and delivery of food to small intestine. Formation of chyme (liquefied food). Some protein digestion. Acid kills bacteria
Duodenum	Digestion of starch, protein and fat
Associated organs: Liver	Production of bile which emulsifies (breaks up) fat
Pancreas	Production of pancreatic juice which contains enzymes for digesting starch, protein and fat
Ileum	Completes digestion of carbohydrates, proteins and fats. Food absorption occurs here
Caecum	No particular function in man
Appendix	No particular function in man
Colon	Water absorption
Rectum	Formation and storage of faeces (solid waste)
Anus	Egestion

Questions on events in different parts of the gut are fairly common. For example:

Q. Describe four important processes concerned with digestion that take place in the buccal cavity.

[CAM]

Mouth (buccal cavity)

Food is **ingested** through the mouth, where the teeth play a large part in preparing the food for digestion. **Mastication** (chewing) is the mechanical breakdown of food by the teeth, aided by the tongue and jaw muscles. The food is broken into small pieces thus exposing a larger surface area for the action of digestive enzymes.

Saliva, secreted by salivary glands
1. lubricates the food for swallowing
2. contains the enzyme salivary amylase (ptyalin) which digests starch to maltose
3. dissolves soluble food.

The teeth and tongue mould the food into a ball (bolus) which is pushed backwards against the roof of the mouth. This sets off the swallowing reflex whereby the bolus is pushed into the oesophagus by the contraction of throat muscles. At the same time, the soft palate closes over the nasal passages and the epiglottis closes over the entrance of the windpipe (trachea) thus stopping food 'going down the wrong way'.

Oesophagus

Inside the oesophagus, the bolus is squeezed down to the stomach by peristalsis (see Fig. 5.12).

Stomach

The stomach is a muscular bag. Its main functions are:
1. To store food temporarily and deliver it, a little at a time, to the duodenum. This prevents the digestive system becoming overloaded with too much food at once.
2. Muscular contractions of the stomach wall churn the solid food into a liquid called **chyme**.
3. The digestion of protein begins here.

The lining of the stomach contains gastric glands which secrete a number of substances:
1. **Hydrochloric acid**, which provides the correct pH for pepsin action (see 2), stops activity of the enzyme salivary amylase, and kills bacteria.
2. The enzyme **pepsin** which digests proteins to polypeptides and peptides.
3. **Rennin**, which is valuable to infants living on milk. It clots milk, so keeping it in the stomach where pepsin can break it down.

At intervals, a ring of muscle, the pyloric sphincter, at the lower end of the stomach relaxes, allowing small quantities of chyme through to the duodenum.

Duodenum

The duodenum is the first 30 cm (12 inches) of the small intestine.

Bile is a green fluid produced by the liver and stored in the gall bladder. It empties into the duodenum along the bile duct (see Fig. 5.11). Bile contains sodium bicarbonate, which neutralizes stomach acid, and bile salts, which emulsify fats (breaks them down into small droplets). Bile does *not* contain any enzymes.

Pancreatic juice, produced by the pancreas, enters the duodenum along the pancreatic duct. It contains sodium bicarbonate to neutralize stomach acidity, as well as three enzymes which together digest the three main types of organic food – carbohydrates, proteins and fats:

$$\text{Starch} \xrightarrow{\text{pancreatic amylase}} \text{maltose}$$

$$\text{Polypeptides} \xrightarrow{\text{trypsin}} \text{peptides + amino acids}$$

$$\text{Fats} \xrightarrow{\text{lipase}} \text{fatty acids + glycerol}$$

Ileum

The second part of the small intestine, the ileum, is where digestion is completed and absorption occurs. Here, the pancreatic enzymes are still active. In addition the glands in the ileum lining secrete enzymes which digest the three main types of organic food down to their soluble constituents:

$$\text{Sucrose} \xrightarrow{\text{sucrase}} \text{glucose + fructose}$$

$$\text{Maltose} \xrightarrow{\text{maltase}} \text{glucose}$$

$$\text{Lactose} \xrightarrow{\text{lactase}} \text{glucose + galactose}$$

$$\text{(disaccharides)} \qquad \text{(monosaccharides)}$$

$$\text{peptides} \xrightarrow{\substack{\text{erepsin (a mixture} \\ \text{of enzymes)}}} \text{amino acids}$$

$$\text{Fats} \xrightarrow{\text{lipase}} \text{fatty acids + glycerol}$$

The organic molecules which were originally large and complex are now small and soluble and can be absorbed.

The lining of the ileum is folded into finger-like projections called **villi** (singular: villus). These greatly increase the surface area for absorption. Other features of the villi which aid absorption are shown in Fig. 5.13.

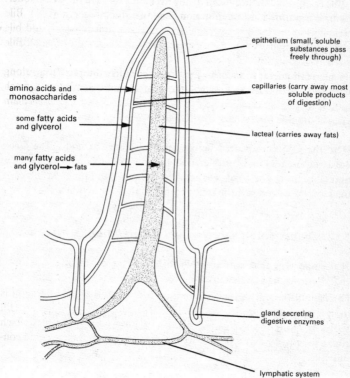

Fig. 5.13. *A villus from the ileum lining*

The final products of digestion are taken into the bloodstream across the lining of the villi. Absorption is by a combination of passive diffusion and active uptake (see page 61). Some fatty acids and glycerol recombine in the intestinal epithelium and as fats enter the lacteals in the centre of the villi (see Fig. 5.13). They then travel through the lymphatic system (page 146) before entering the bloodstream. All other products of digestion travel in the bloodstream to the liver (page 96) where their rate of supply to body tissues is regulated. The digested food is distributed round the body in the bloodstream and is delivered to cells via the tissue fluid (page 144). All cells in the body require a supply of organic foods which are finally **assimilated** when they enter the cell.

Caecum and appendix

In man, these are reduced in size and have no special function.

Colon (large intestine)

Its main function is to absorb water from the fluid contents. The resulting semi-solid mass (**faeces**) is passed into the rectum by peristalsis.

Rectum

Here the faeces are stored before removal from the body. The faeces are composed of indigestible food (mainly cellulose and other plant fibres), bacteria and dead cells scraped from the gut lining. At intervals, the faeces are **egested** through the anus by muscular action.

5.12. *Digestion, Absorption and Assimilation*

It is important that you are able to follow the chain of events, from ingestion through to assimilation, for the major types of organic food. For example:

Q. Describe the parts played by the following in the digestion and absorption of protein in the diet of a mammal:
(*i*)   *the stomach*
(*ii*) *the pancreas*
(*iii*) *the small intestine.*   [9]
[SUJB]

Q. Name and give the action of the various enzymes in the digestion of carbohydrates in a mammal.   [10]
[LON]

Carbohydrates

Carbohydrates are ingested as monosaccharides, disaccharides or polysaccharides. The last two need to be digested to monosaccharides before absorption.

Digestion

In the mouth (buccal cavity):

$$\text{starch} \xrightarrow[\text{(ptyalin)}]{\text{enzyme salivary amylase}} \text{maltose}$$

In the stomach: no carbohydrate digestion.
In the duodenum:

$$\text{starch} \xrightarrow{\text{enzyme pancreatic amylase}} \text{maltose}$$

In the ileum:

$$\text{maltose} \xrightarrow{\text{maltase}} \text{glucose}$$

$$\text{sucrose} \xrightarrow{\text{sucrase}} \text{glucose} + \text{fructose}$$

$$\begin{array}{c}\text{lactose} \\ \text{(disaccharides)}\end{array} \xrightarrow[\substack{\text{enzymes secreted by} \\ \text{lining of ileum}}]{\text{lactase}} \begin{array}{c}\text{glucose} + \text{galactose} \\ \text{(monosaccharides)}\end{array}$$

All digestible carbohydrates are now monosaccharides and can be absorbed across the lining of the ileum.

Absorption

In the ileum, monosaccharides (mainly glucose) pass through the epithelium of villi, through capillary walls and into the blood plasma. Absorption is by a combination of passive diffusion and active uptake. The glucose then travels in the hepatic portal vein to the liver.

Assimilation

What happens to glucose on reaching the liver is determined by the blood glucose level.
1. Glucose is the body's main source of energy. All cells need a supply of glucose for respiration (see page 105).

2. Excess glucose is converted to glycogen in the liver and skeletal muscles. When the blood glucose level drops, the liver converts its glycogen back into glucose and releases it into the bloodstream. In this way, the blood glucose level is kept relatively constant.

3. Glucose in large excess is converted to fat and stored in adipose tissue beneath the skin (page 174) and around certain internal organs. Hence, a large amount of carbohydrate in the diet may lead to obesity. The fat may be respired later (see page 96).

Proteins

Proteins are ingested in the form of polypeptide chains of varying length. These have to be broken down to amino acids before absorption.

Digestion
  In the mouth: no protein digestion
  In the stomach:

$$\text{proteins} \xrightarrow{\text{enzyme pepsin}} \text{polypeptides}$$

$$\text{milk protein} \xrightarrow{\text{rennin*}} \text{clotted protein}$$

  In the duodenum:

$$\text{trypsinogen (inactive enzyme)}$$
$$\downarrow$$
$$\text{trypsin   (active enzyme)}$$
$$\downarrow$$

$$\text{proteins + polypeptides} \longrightarrow \text{peptides + amino acids}$$
$$\text{(in pancreatic juice)}$$

  In the ileum (trypsin still active):

$$\text{peptides} \xrightarrow[\substack{\text{secreted by the lining} \\ \text{of the ileum)}}]{\text{erepsin (several enzymes}} \text{amino acids}$$

All proteins have now been digested to amino acids which can be absorbed across the lining of the ileum.

---

* In children only. Milk is clotted so that it remains in the stomach where it can be digested by pepsin.

Absorption

As with monosaccharides, amino acids pass into capillaries in the ileum wall by a combination of passive and active processes. The amino acids then travel in the hepatic portal vein to the liver.

Assimilation

What happens to amino acids on reaching the liver depends on the body's requirement for protein.

1. All cells in the body need a supply of amino acids to build up their own proteins.

2. Excess amino acids are *not* stored, but in the liver are **deaminated** (their amine group removed). The amine group (NH_2) is altered to the very poisonous ammonia (NH_3) which is then converted to the less harmful substance **urea**. Urea is released into the bloodstream, transported in the blood to the kidneys, and then excreted in the urine (see page 169).

The remainder of the amino acid molecule is broken down in the liver to release energy in respiration.

Fats

Fats, unlike carbohydrates or proteins, do not readily mix with water but tend to stay clumped together as globules. Before fats can be effectively digested they must be physically broken down into smaller droplets (emulsification). This increases their surface area exposed to digestive enzymes.

Digestion

In the mouth: no fat digestion.

In the stomach: no fat digestion.

In the duodenum:

$$\text{fat globules} \xrightarrow{\text{bile salts}} \text{fat droplets}$$

$$\text{fat} \xrightarrow[\text{pancreatic juice}]{\text{lipase in}} \text{fatty acids} + \text{glycerol}$$

In the ileum (pancreatic lipase still active):

$$\text{fats} \xrightarrow[\text{lining of ileum}]{\text{lipase secreted by}} \text{fatty acids} + \text{glycerol}$$

Absorption

Only some of the fatty acids and glycerol enter the capillaries of the villi inside the ileum. A large proportion join together again in the intestinal epithelium and pass as fats into the **lacteals**. The fluid in the lacteals enters the lymphatic system which eventually empties its contents into the bloodstream (page 146). Thus, most of the products of fat digestion bypass the liver.

Assimilation

1. All cells in the body need fats to construct and repair membranes in the cell.
2. Fats are an important source of energy released by respiration.
3. Fats not immediately broken down in respiration are stored in adipose tissue under the skin and around certain internal organs. When required for energy, the stored fats are sent to the liver where they are converted to substances which can be readily respired.

5.13. *The Liver*

The liver is the largest organ in the body. It has over 500 functions. It is reddish-brown and lies at the top of the abdomen just below the diaphragm. The liver has the following connections with other parts of the body:

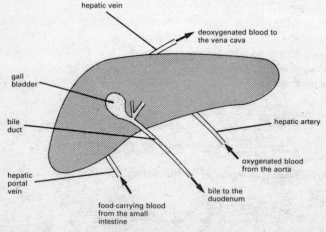

Fig. 5.14. *Connections to and from the liver*

The **hepatic artery** brings oxygenated blood from the aorta while the **hepatic portal vein** brings absorbed food from the alimentary canal.

All blood leaving the liver travels along the **hepatic vein** to the vena cava. Bile, which is stored in the gall bladder, empties into the duodenum along the **bile duct**.

The liver is a major homeostatic organ, i.e. it helps keep conditions within the body constant. Dissolved food substances absorbed through the gut wall are delivered in the blood straight to the liver. The liver then controls the 'through-flow' of these substances according to the body's needs.

Functions of the liver

The ten main functions of the liver are more easily remembered if grouped into categories.

Three functions concern the major types of organic foods – carbohydrates, proteins and fats:

1. The liver controls blood glucose level (see page 94).
2. It deaminates excess amino acids (see page 95).
3. It controls the body's use of fats (see page 96).
    It has two manufacturing (synthetic) functions:
4. It forms plasma proteins, e.g. fibrinogen (see page 135).
5. It forms bile (see page 90).
    It has two storage functions:
6. It stores the fat-soluble vitamins A and D and the water soluble vitamin B_{12}.
7. It stores iron from the breakdown of haemoglobin (see 8 below).
    It has two 'breaking-down' functions:
8. Old red blood cells are removed from the circulation and their haemoglobin broken down by the liver.
9. Detoxification. It converts poisonous substances to harmless ones.
    One heat-related function:
10. Being such a large, active organ, respiration in the liver is the main source of body heat.

5.14. *Teeth*

Teeth are produced by the skin covering the jaws and are set into sockets in the jawbone.

The number and arrangement of teeth in the jaws is called the **dentition**.

The teeth of mammals differ from those of other vertebrates:

Mammals	*Other vertebrates*
Teeth have a **pulp cavity**	No pulp cavity
Teeth specialized for different functions (**heterodont** dentition)	Teeth all of the same type (**homodont** dentition)

In mammals, a set of **milk teeth** develop first and are later pushed out and replaced by **adult (permanent)** teeth.

Mammalian teeth are layers of modified bone tissue, supplied with food and oxygen by the pulp cavity (see Fig. 5.15), and shaped according to function.

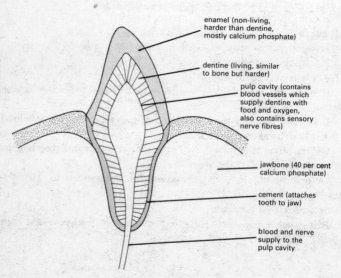

enamel (non-living, harder than dentine, mostly calcium phosphate)

dentine (living, similar to bone but harder)

pulp cavity (contains blood vessels which supply dentine with food and oxygen, also contains sensory nerve fibres)

jawbone (40 per cent calcium phosphate)

cement (attaches tooth to jaw)

blood and nerve supply to the pulp cavity

Fig. 5.15. *V.S. of canine tooth of a mammal*

There are four types of teeth in mammals:

Type	Position in the jaw	Appearance	Function
Incisor	At the front	Chisel-shaped, single root	Biting or gnawing
Canine	Behind the incisors	Pointed, single root	Stabbing or tearing
Premolar	Behind the canines	Flattened with 2 cusps, 1 or 2 roots	Grinding, crushing or slicing
Molar	Behind the premolars	Flattened with 4 cusps, 2 or 3 roots	" "

(Premolar and Molar grouped as "(cheek teeth)")

The number and type of teeth are determined by the mammal's diet.

1. **Man** is an **omnivore** and has a 'generalized dentition':
(a) The incisors are flat for biting all types of food.
(b) Canines are present but are not large.
(c) The cheek teeth have rounded cusps for chewing a mixed diet.

2. A **dog** is a **carnivore**. Its dentition is adapted for seizing and killing prey, tearing and chewing flesh, and cracking and crushing bones. Refer to Fig. 5.16 as you go through the following points:

Fig. 5.16. *Comparison of herbivore and carnivore jaws and teeth*

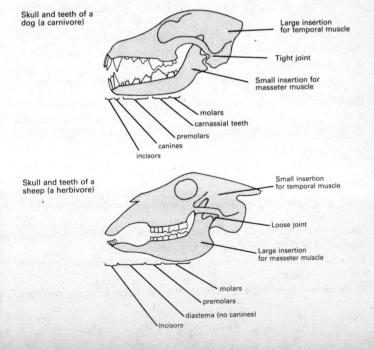

Skull and teeth of a dog (a carnivore)
Large insertion for temporal muscle
Tight joint
Small insertion for masseter muscle
molars
carnassial teeth
premolars
canines
incisors

Skull and teeth of a sheep (a herbivore)
Small insertion for temporal muscle
Loose joint
Large insertion for masseter muscle
molars
premolars
diastema (no canines)
incisors

(a) Pointed incisors are used for nibbling meat from bones.

(b) Large, pointed canines are used to kill the prey, by piercing veins in the neck.

(c) Premolars and molars are flattened along the line of the jaw and have a scissor-like action for slicing flesh and crunching bones. The **carnassial teeth** are specially adapted for this. Sliced meat is swallowed without chewing.

(d) The tight joint between skull and jaw and the position of the masseter muscle restricts side to side movement of the jaw.

(e) The temporal muscle (which moves the jaw up and down) is large, producing a powerful bite.

3. A **sheep** is a **herbivore**. Its dentition is specially adapted to crop and grind tough vegetation. Refer to Fig. 5.16 as you go through the following points:

(a) The flattened incisors of the bottom jaw meet against a horny pad on the top jaw and are used for tearing or pulling out grass.

(b) The canines are absent. A large gap, the **diastema**, serves to separate newly cropped grass from that being ground up by the cheek teeth.

(c) The cheek teeth are large and ridged, the ridges on the bottom jaw fitting into those on the top. The two sets of teeth grind against each other in a circular motion.

(d) The cheek teeth grow continuously because of the considerable wear they receive.

(e) The masseter muscle (which moves the jaw from side to side) is large. The loose joint between jaw and skull allows a sideways rocking action as grass is chewed.

(f) The temporal muscle (which moves the jaw up and down) is small.

Q. (a) *Make a labelled diagram to show the structure of a mammalian tooth.*

[8]

(b) *In what ways do teeth of mammals differ from those of other vertebrates?*

[2]

[OXF]

5.15. *The Alimentary Canal of Carnivores and Herbivores*

Just as the dentition and jaw structure of mammals differ according to diet, so the basic plan of the gut is modified.

1. **Carnivores** have a relatively short gut because their food is protein-rich and easy to digest.

2. **Herbivores** have a diet rich in plant fibre (indigestible to omnivores

and carnivores) and need special adaptations to utilize this source of food. Mammals cannot make their own cellulose-digesting enzymes, but herbivorous mammals have developed an association (**symbiosis**) with certain micro-organisms which provide the necessary enzymes. A herbivore's intestines are long as plant material is difficult to digest.

(a) **Ruminants**, e.g. cows and deer, have a large outgrowth of the stomach, the **rumen**. Here, symbiotic bacteria produce cellulose-digesting enzymes. The freshly ingested food is passed to the rumen where it is 'fermented' (partially digested). It is later regurgitated and thoroughly chewed ('chewing the cud') before passing into the true stomach.

(b) **Non-ruminants**, e.g. horses and rabbits, do not have a rumen, but culture their cellulose-digesting bacteria in the caecum and appendix. Rabbits 'refect', eat their faeces, thus passing food through the gut twice.

All these adaptations of herbivores serve to increase their ability to digest fibre-rich food.

5.16. *Feeding Methods in Holozoic Organisms*

On A E B and J M B courses you need to be familiar with feeding methods in a range of animals.

Most animals show **extracellular digestion**: food is digested outside cells, in a gut or outside the organism.

Single-celled animals, e.g. *Amoeba* (page 326) show **intracellular digestion**: food is digested within the cell.

The coelenterate, *Hydra*, shows both intra- and extracellular digestion (see page 329).

In general, animals obtain their food in one of three forms:
1. as large solids
2. as solids suspended in liquid
3. as liquids.

1. **Large solids** must be broken down by biting and chewing into particles which are small enough to be ingested and swallowed.

Those insects (page 335) which have chewing mouthparts include the locust (a herbivore), the cockroach (an omnivore), and the ground beetle (a carnivore).

Those mammals (page 342) which have chewing mouthparts include the sheep (a herbivore), man (an omnivore) and the dog (a carnivore).

2. **Suspended solids** need to be filtered out of the liquid before they can be ingested. Usually a sieve-like apparatus is used to do this. As suspended organisms are usually very small, vast quantities of water must be filtered

to provide an adequate food supply. This method of feeding (filter-feeding) is found in certain aquatic organisms, such as the herring (page 337).

3. Those animals which feed on **liquids** have sucking mouthparts. In the case of the housefly, the mouthparts form a hollow tube (the **proboscis**) which is used for releasing saliva as well as sucking up food. The saliva contains digestive enzymes which convert solid organic matter to a semi-digested liquid. This liquid can then be drawn up the proboscis and into the stomach by muscular contraction.

The mouthparts of the mosquito are for piercing as well as sucking. The proboscis is pointed and in the female is used to pierce mammalian skin to obtain blood, and in the male, to pierce plants to obtain phloem.

5.17. *Saprophytic Nutrition*

Saprophytes play a major role in the cycling of chemicals in nature; see decay bacteria and fungi (Figs 14.1 and 14.2, pages 289–90) and nitrogen-fixing and nitrifying bacteria (page 309 and Fig. 14.2). *Mucor* (page 314) is an example of a saprophytic fungus.

5.18. *Parasitic Nutrition*

For examples of parasites, see viruses (page 305), disease-causing bacteria (page 308), the fungus *Phytophthora* (page 315) and the tapeworm (page 330).

Definitions

Ingestion
Digestion
Absorption
Assimilation
Egestion
Secretion

Key Words

Autotrophic
Heterotrophic
Holozoic
Saprophytic
Parasitic
Host
Chloroplast
Midrib
Vein
Lamina
Palisade mesophyll
Spongy mesophyll
Air space
Epidermis
Cuticle
Guard cell
Stomata
Xylem
Phloem
Variegated
Light reaction
Dark reaction
Limiting factor
Wilting
Water culture

Carnivore
Herbivore
Omnivore
Balanced diet
Peristalsis
Mucus
Buccal cavity
Oesophagus
Stomach
Duodenum
Liver
Pancreas
Ileum
Caecum
Appendix
Colon
Rectum
Anus
Salivary amylase
Bolus
Hydrochloric acid
Pepsin
Rennin
Chyme
Pyloric sphincter

Emulsify
Lipase
Trypsin
Erepsin
Sucrase
Maltase
Lactase
Villi
Hepatic artery
Hepatic vein
Bile duct
Gall bladder
Pulp cavity
Homodont
Heterodont
Incisor
Canine
Premolar
Molar
Diastema
Carnassial teeth
Extracellular
    digestion
Intracellular digestion

Exam Questions

1. (a) Give a balanced equation which summarizes the process of photo-synthesis. [5]
(b) Describe in detail an experiment which would demonstrate that light is needed for photosynthesis. [13]
(c) What is the importance of photosynthesis to animals? [7]
[LON]
2. State four environmental conditions that enable photosynthesis to occur [4]
[O&C]

3. *For each of the following food materials describe* one *test you would use to show its presence:*
(*i*)   *a reducing sugar*
(*ii*) *starch*
(*iii*) *a protein*
(*iv*) *a fat*   [15]
[OXF]

4. (*a*) *List the food requirements of mammals.*   [7]
(*b*) *In what ways are herbivorous and carnivorous mammals adapted to their respective diets?*   [10]
(*c*) *How do*
(*i*) *a named protozoan*
(*ii*) *tapeworms, obtain their supplies of nourishment?*   [8]
[LON]

(For section (c) of the above question see pages 326 and 330.)

5. *The most important function of the villi lining the small intestine of a mammal is to*
A. *increase the surface area*
B. *push the food along*
C. *secrete increased quantities of enzymes*
D. *trap bacteria*
E. *direct the flow of blood.*   [1]
[LON]

6. *Vitamin A is essential in man's diet to prevent*
A. *pellagra*
B. *scurvy*
C. *beri-beri*
D. *rickets*
E. *night-blindness.*   [1]
[LON]

6. Respiration, Gas Exchange and Breathing

Definition: Respiration *is the breakdown of organic substances, within cells, to release energy.*

Respiration occurs in all living cells since every cell needs a supply of energy to remain alive.

The term respiration should not be confused with gas exchange or breathing (ventilation).

Definition: Gas exchange *is the exchange of oxygen for carbon dioxide between an organism and its surroundings.*

Gas exchange usually accompanies respiration. The surface at which it occurs is called a respiratory or gas exchange surface.

Breathing or ventilation refers to the movements which cause a renewal of air or water at a respiratory surface. Breathing occurs in larger animals such as the larger arthropods and in vertebrates.

Definition: Breathing (ventilation) *is the maintenance of a flow of air or water over a respiratory surface.*

The term respiration is sometimes referred to as **internal**, **cell** or **tissue respiration**, and gas exchange and breathing grouped under the term **external respiration**. This use of terms can lead to confusion. The more modern naming system used in this book is simpler and clearer.

You must be absolutely clear about these terms. They often form the basis of long-answer questions. Clearly if you have got their meaning wrong you are in trouble.

Q. Distinguish between respiration and ventilation (breathing). [6]
[OXF]

In its simplest form respiration is represented by the chemical equation:

$$\text{glucose} + \text{oxygen} \rightarrow \text{carbon dioxide} + \text{water} + \text{energy}$$
$$C_6H_{12}O_6 + 6O_2 \rightarrow 6CO_2 + H_2O + \text{energy}$$

As with photosynthesis (page 64), this represents a simplification of a complex series of reactions.

The key feature of respiration is that it releases energy. Respiration is a characteristic feature of life and all organisms need a continual supply of energy to carry out essential processes. These processes include:

1. Anabolic metabolism (page 35) in which energy is used in chemical synthesis, i.e. building up complex substances from simpler ones.
2. Muscle contraction to do mechanical work (page 224).
3. Generation and transmission of nerve impulses (page 183).
4. Active transport (page 61) of substances into and out of cells.
5. In birds an mammals, heat energy is used to maintain a high body temperature (page 175).

Energy is the capacity to do work. It exists in a variety of forms such as heat, light, electrical, chemical and mechanical energy. Energy cannot be created or destroyed, but it can be changed from one form into another.

In respiration, chemical energy is converted into heat energy and usable chemical energy. The chemical energy comes from the bonds which link atoms in the glucose molecule. When these bonds are broken, energy is released. Some is given off as heat and some is trapped and stored as usable chemical energy within a substance called adenosine triphosphate (ATP for short). The ATP then delivers the energy to those parts of the cell which require it.

The chemical energy within the glucose molecule originally came from the sun as light energy. Plants are able to trap this and convert it into chemical energy during **photosynthesis**:

$$\text{carbon dioxide} + \text{water} \xrightarrow[\text{chlorophyll}]{\overset{\text{light energy}}{\text{(sunlight)}}} \text{glucose} + \text{oxygen}$$

$$6CO_2 + 6H_2O \longrightarrow \underset{\substack{\text{(store of} \\ \text{chemical energy)}}}{C_6H_{12}O_6 + 6O_2}$$

When the chemical bonds within glucose are broken, energy is released:

$$C_6H_{12}O_6 + 6O_2 \rightarrow 6CO_2 + 6H_2O + \text{energy}$$

As we noted earlier (page 78) respiration is essentially the reverse of photosynthesis. Photosynthesis stores energy while respiration releases it.

$$\text{carbon dioxide} + \text{water} \underset{\underset{\text{heat and chemical energy}}{\text{RESPIRATION}}}{\overset{\overset{\text{light energy}}{\text{PHOTOSYNTHESIS}}}{\rightleftharpoons}} \text{glucose} + \text{oxygen}$$

Only green plants photosynthesize, but *all* organisms respire. Thus, green plants provide the energy for all forms of life.

6.1. *Aerobic and Anaerobic Respiration*

There are two forms or types of respiration, **aerobic** and **anaerobic**. Aerobic respiration requires oxygen, anaerobic respiration does not.

Aerobic respiration is the much more efficient process and releases much more energy from each glucose molecule. The reason is, in aerobic respiration the glucose is broken down into very simple substances, whereas in anaerobic respiration the products are still complex and energy-rich (contain large amounts of stored energy).

The key features of aerobic and anaerobic respiration are summarized in Table 6.1. Do not worry if you cannot remember the chemical formulae; the names and energy output are more important.

Table 6.1. *Aerobic and anaerobic respiration compared*

Aerobic respiration	Anaerobic respiration
1. *most organisms* $C_6H_{12}O_6 + 6O_2 \rightarrow$ glucose oxygen $6CO_2 + 6H_2O + 2{,}880\ kJ$ carbon water energy dioxide	1. *in plants* (certain fungi, bacteria and to a limited degree in higher plants during periods of oxygen shortage) $C_6H_{12}O_6 \rightarrow$ glucose $2C_2H_5OH + 2CO_2 + 210\ kJ$ ethanol carbon energy dioxide (2-carbon compound) *in animals* (tapeworm and to a limited extent in mammalian skeletal muscle during vigorous exercise) $C_6H_{12}O_6 \rightarrow$ glucose $2CH_3CH(OH)COOH + 150\ kJ$ lactic acid energy (3-carbon compound)
2. Requires oxygen	2. Does not require oxygen
3. Within the cell, the reactions take place in mitochondria (page 40)	3. Within the cell, the reactions take place 'free' in the cytoplasm
4. All or most of the available energy within the glucose molecule is released	4. Only part of the energy within the glucose molecule is released
5. Water is a product of the reaction	5. Water is not a product of the reaction

Q. *Give three differences between aerobic and anaerobic respiration.*   [3]
[OXF]

Aerobic and anaerobic respiration are not two completely different reactions, rather they are two different stages in a sequence of reactions.

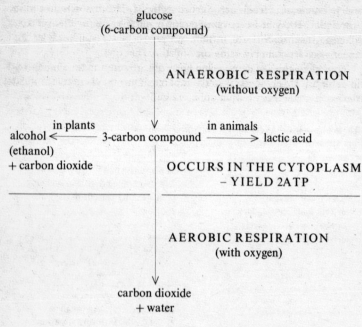

glucose
(6-carbon compound)

ANAEROBIC RESPIRATION
(without oxygen)

in plants   in animals
alcohol ⟵⟶ 3-carbon compound ⟶ lactic acid
(ethanol)
+ carbon dioxide

**OCCURS IN THE CYTOPLASM
– YIELD 2ATP**

AEROBIC RESPIRATION
(with oxygen)

carbon dioxide
+ water

**OCCURS IN MITOCHONDRIA
– YIELD 36ATP**

TOTAL YIELD 38ATP

Anaerobic respiration

Anaerobic respiration, the first stage of respiration, does not require oxygen but releases only a small proportion of the available energy in glucose. In those organisms which live in an environment without oxygen, e.g. certain bacteria (page 307) and in the tapeworm (page 330), the reaction proceeds only as far as this point. In plants, alcohol and carbon dioxide are formed, and in animals, lactic acid. **Fermentation** is the term used to describe anaerobic respiration in organisms which derive most of their energy from this process.

Certain organisms such as the yeast, a fungus (page 304), can readily switch from anaerobic to aerobic respiration. Yeast's anaerobic ability is used by man in beer- and wine-making (in producing alcohol) and in bread-making (the carbon dioxide causes bread to rise).

Most animals have only a limited ability to sustain anaerobic respiration. In mammals, certain organs such as the brain and heart must have a continual supply of oxygen for aerobic respiration and without it they will soon die (brain death in man occurs within 2–3 minutes of the oxygenated blood supply being cut off).

Certain tissues in the mammalian body can, however, maintain anaerobic respiration for short periods of time. During vigorous physical exercise, e.g. running or swimming, not enough oxygen may reach contracting skeletal muscle (page 221) to supply all the tissues' energy needs. In this case, aerobic respiration is supplemented by anaerobic respiration. This state of affairs will only exist for a short time because the products of anaerobic respiration (lactic acid) accumulate and cause fatigue.

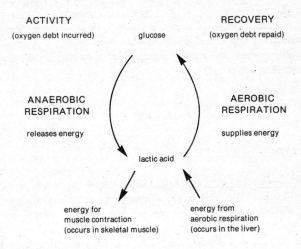

During the period of anaerobic respiration an 'oxygen debt' is built up. Later, aerobic respiration (requiring oxygen) will supply the energy to convert the lactic acid back to glucose. Anaerobic respiration is called upon at times of high energy demand (and oxygen shortage) and the energy it releases is paid back when oxygen is available.

After accumulating in the muscle, the lactic acid enters the bloodstream and passes to the liver. Here it is converted to glucose when the oxygen debt is repaid. The glucose is then returned to the muscle and stored as glycogen.

The 'oxygen debt' is the reason why it is necessary to go on breathing heavily for a short time after strenuous exercise; the 'oxygen debt' is being paid off.

Aerobic respiration

In the presence of oxygen, most organisms will continue the breakdown of glucose using the second series of reactions which occur in mitochondria. Using oxygen, glucose is oxidized to the simple molecules carbon dioxide and water and the maximum amount of energy extracted. The energy yield for the second series of reactions (aerobic respiration) is nearly twenty times greater than for anaerobic respiration.

Q. (*a*) *In the process of respiration living organisms can release and use chemical energy within a carbohydrate molecule.*
(*i*)   *Write an equation, in words or symbols, to show the beginning and end products of respiration when oxygen is freely available.*
(*ii*) *Write an equation, in words or symbols, to show the beginning and end products of respiration in one situation where oxygen is not available.*
(*iii*) *What happens in active muscle tissue when the oxygen supply is temporarily insufficient?*

[AEB]

6.2. *ATP**

Of the energy released in respiration reactions, 80 per cent is given off as heat which has little value to most organisms except homoiotherms (see page 175). The remaining 20 per cent is usable chemical energy. Inside the cell, this energy is trapped as a temporary energy store inside the substance adenosine triphosphate, ATP. This chemical then delivers the stored energy to any reaction or process in the cell which requires it. This is the case in *all* organisms. ATP carries energy from the respiration reactions that release it to those reactions in the cell that require it.

The ATP molecule contains three phosphate groups, and it stores energy in the chemical bond between the middle and outer phosphate groups. This bond is formed, using the energy from respiration, by adding

* This is not required on all syllabuses.

a phosphate group to a molecule of the substance, adenosine diphosphate, ADP:

$$A\text{-}P\text{-}P + \quad P \quad + \quad energy \quad \rightarrow A\text{-}P\text{-}P \sim P$$

ADP   phosphate   from respiration   ATP

The bond so formed is called a high-energy bond because of the large amount of stored energy it contains. When broken by hydrolysis, the energy is released, and ADP is re-formed:

$$A\text{-}P\text{-}P \sim P \quad \rightarrow \quad A\text{-}P\text{-}P \quad + \quad P \quad + \quad energy\ to\ do\ work$$

ATP   ADP   phosphate   in the cell

In this way ATP is recycled within the cell. It is built up using energy released from the breakdown of glucose, and then broken down to supply energy to any process in the cell that requires it. Each ATP molecule represents a 'packet' of available energy.

energy from respiration

$$ADP + P \rightleftharpoons ATP$$

energy to do work in
the cell

6.3. *Respiration Experiments*

A demonstration of respiration in material is one indication that the material is living.

Measurements of the rate of respiration give some idea of the rate of metabolic activity.

If we look at the equation for respiration on page 105 we see that an organism respiring aerobically should:

1. use up carbohydrate
2. use up oxygen
3. give off carbon dioxide
4. produce water or water vapour
5. release energy.

If we can show that one of the above is occurring we can infer that the organism is living and respiring. In practice, 1, 2, 3 and 5 are suitable tests for respiration. 4 is not an effective test because:

(a) Non-living as well as living material may give off water by evaporation.

(b) Only a fraction of the water given off by an organism comes from respiration; most has originally been absorbed.

Of the four possibilities remaining, 3 and 5 are the easiest to demonstrate, and questions on these respiration experiments are commonly included in exams.

Investigation of carbon dioxide production by living organisms

This experiment investigates whether living organisms release carbon dioxide.

Method

The apparatus is shown in Fig. 6.1 (over). The experiment is effective using small animals or, over a longer period, using plant material.

Potassium or sodium hydroxide in flask A is used to remove carbon dioxide from incoming air. The lime water in flask B confirms that the carbon dioxide has been removed (if not, the lime water turns milky). The air delivered to the organism in the specimen chamber is free of carbon dioxide. If carbon dioxide is found in the air leaving the chamber (as shown by the lime water in flask C turning milky), then it must have come from the organism. What would be an effective control for this experiment?

If a green plant is used as the test organism, then certain precautions must be taken (see Fig. 6.1):

1. The vessel must be 'blacked out' to prevent photosynthesis occurring.
2. If a potted plant is used, the soil must be enclosed in carbon dioxide-proof material so that respiration of organisms in the soil does not affect the result.

Results

If a live organism is placed in the specimen chamber, the lime water in flask C will eventually turn milky. The time taken for this to occur can be used as a rough indication of the rate of carbon dioxide release. (Weight for weight, live animal material releases much more carbon dioxide than plant material. Why do you think this is so?)

If an empty specimen chamber is used as the control, it is found that the lime water in flask C does not turn milky.

Conclusions

Living material releases carbon dioxide. We can assume the carbon dioxide to have come from respiration, although we do not know this to be the case from the experiment.

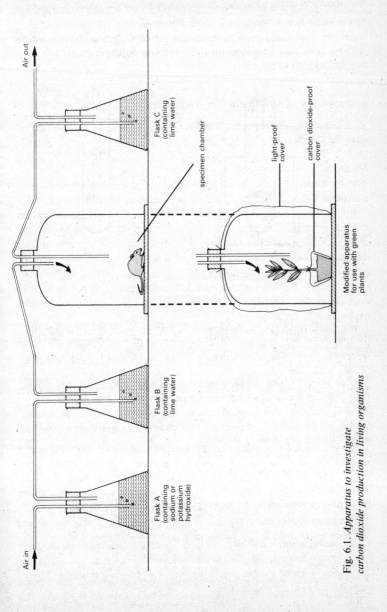

Fig. 6.1. *Apparatus to investigate carbon dioxide production in living organisms*

Air in

Flask A
(containing
sodium or
potassium
hydroxide)

Flask B
(containing
lime water)

Flask C
(containing
lime water)

Air out

specimen chamber

light-proof cover

carbon dioxide-proof cover

Modified apparatus
for use with green
plants

Investigation of carbon dioxide production by anaerobic organisms

A modification of the above experiment can be used to show that anaerobic organisms such as yeast can produce carbon dioxide in the absence of oxygen. The experiment is set up as shown in Fig. 6.2.

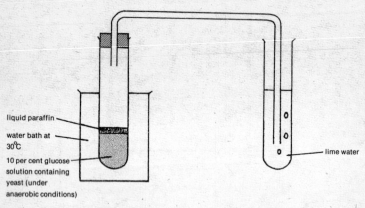

Fig. 6.2. *Apparatus to investigate carbon dioxide production under anaerobic conditions*

The water is boiled to remove carbon dioxide before the yeast and glucose are added. The layer of liquid paraffin prevents oxygen entering the liquid. A control apparatus, identical but without the yeast, is also set up.

After several hours the lime water in the test apparatus turns milky, while that in the control apparatus remains clear. The results, taken together, indicate that yeast liberates carbon dioxide under anaerobic conditions. In addition, the presence of alcohol (a product of anaerobic respiration) can be detected by smell.

Investigation of energy release in germinating seeds

This experiment investigates whether live seeds release energy. Heat production is a good indication of energy release.

Method

A suitable experiment is shown in Fig. 6.3 (over). Wheat seeds are used as the experimental material. Two batches of seeds of equal weight are used. One batch is killed by placing in boiling water for 10 minutes. These seeds will be used for the control.

Both batches are rinsed with hypochlorite solution to prevent bacterial or fungal growth. The live seeds are then placed in flask A and the dead seeds in flask B. Thermometers are inserted and the mouths of the flasks plugged. The temperature within the flasks is recorded at one-day intervals.

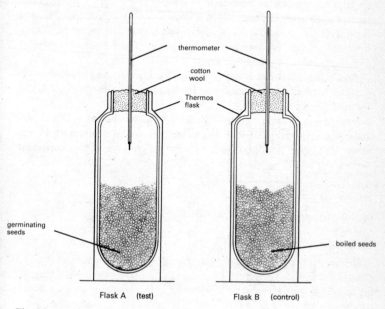

thermometer

cotton wool

Thermos flask

germinating seeds

boiled seeds

Flask A   (test)

Flask B   (control)

Fig. 6.3. *Apparatus to investigate heat production in germinating seeds*

Results
Typical results are as follows:

| | Temperature °C | |
Day	Flask A	Flask B
1	21	20
2	23	20
3	25	19
4	26	20

Conclusion
The results indicate a temperature rise in flask A, whereas the temperature in flask B remains nearly constant. Heat is released by the live seeds in flask A but not by the dead seeds in flask B. The conclusion is that living seeds release heat. This can be assumed to have come from respiration although we do not know this to be the case from this experiment.

Investigation of gas (oxygen) uptake by living organisms

The experiment investigates whether living organisms take up oxygen.

Method

Oxygen uptake cannot be measured directly. However, if an organism is respiring aerobically the oxygen it takes up will be converted to carbon dioxide. If the carbon dioxide is absorbed into soda lime or potassium hydroxide in a closed container, any drop in gas volume can be attributed to oxygen uptake by the organism.

A suitable apparatus is shown in Fig. 6.4. Germinating seeds or small animals are suitable test material. In the case of seeds, dead seeds can be used for the control. The apparatus is immersed in a water bath to minimize temperature changes which might affect the gas volume readings. The apparatus is left in the water bath for five minutes with the screw tops open, and then the screw tops are closed and the experiment begun.

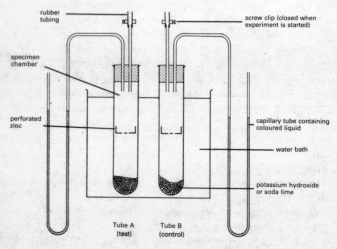

Fig. 6.4. *Apparatus to investigate oxygen uptake in living organisms*

Results

After about half an hour the level of coloured liquid in the capillary tubing of A will typically rise on the side nearest the specimen chamber. In the capillary tubing attached to B little or no change in the level of the liquid is observed.

Conclusions

The results indicate a drop in volume of gas in tube A but little or no change in the volume in tube B. The drop in volume in tube A can be attributed to uptake of oxygen by the live specimen(s). The experiment shows that a living organism takes up oxygen.

NOTE. We have assumed that the gas taken up is oxygen but we have not positively shown this to be the case.

6.4. *Gas Exchange in a Variety of Animals*

All aerobic organisms require oxygen for respiration and produce carbon dioxide as a waste product. At some point on the body surface these gases must be exchanged between the organism and the environment. Oxygen diffuses into the organism and carbon dioxide diffuses out, along a concentration (diffusion) gradient (see page 56).

In small simple organisms such as *Amoeba* (page 326) and *Hydra* (page 328) all parts of the body are only a short distance from the external environment and gas exchange occurs over the entire body surface. As organisms increase in size and complexity the inner parts of the body are too far away from the outside to get their oxygen and get rid of carbon dioxide by diffusion alone. Also the surface area to volume ratio decreases and diffusion cannot proceed at a high enough rate to supply the tissues (see Fig. 6.5). A specialized respiratory surface, e.g. the gill, is required, and a transport system to deliver gases to and from all cells in the body.

All respiratory surfaces have the following features which permit diffusion:

1. They are thin (to allow gases to diffuse rapidly through).
2. They are moist (gases must be in solution before they can pass across a cell membrane).
3. They have a large surface area (to maximize diffusion).
4. They are permeable to gases.

In addition, in larger organisms and those which have a high energy demand, more oxygen is required and the respiratory surface must satisfy the following conditions:

5. It must be well ventilated.
6. It must be richly supplied with blood, which usually contains a respiratory pigment, e.g. haemoglobin. This increases the amount of oxygen that can be carried in the blood.

NOTE. Insects are an exception (see page 120).

You need to be familiar with the gas exchange surfaces of a range

As organisms increase in size, the body surface increases by the square of its dimensions, whereas the volume increases by the cube. The result is that the larger the organism, the smaller the surface area : volume ratio. This can be illustrated by looking at cubes of increasing size:

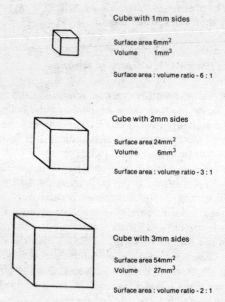

Cube with 1mm sides

Surface area 6mm^2
Volume   1mm^3

Surface area : volume ratio - 6 : 1

Cube with 2mm sides

Surface area 24mm^2
Volume   6mm^3

Surface area : volume ratio - 3 : 1

Cube with 3mm sides

Surface area 54mm^2
Volume   27mm^3

Surface area : volume ratio - 2 : 1

Fig. 6.5. *Surface area to volume ratios*

The effect of this is that in larger organisms there is less surface area to supply each volume of body tissue. A size is soon reached where simple diffusion across the body surface is insufficient to supply the needs of all parts of the body. At this stage, specialized respiratory surfaces and transport systems are required.

Surface area : volume ratios also help explain why cells are so small. As they increase in size, the surface area which supplies food and oxygen to, and removes wastes from, each volume of protoplasm, decreases. A limiting size is soon reached. Beyond this all parts of the cell are unable to obtain supplies and get rid of wastes quickly enough to support life.

of animals of different size and living in different environments. In each case the basic principles are the same. Notice how the above features apply to each example. These are the features you need to point out when answering questions on gas exchange (see, for example, Question 3 on page 131).

Gas exchange in protozoa, e.g. Amoeba (*page 326*)

The gas exchange surface is the cell membrane. It is thin, moist (surrounded by water), has a large surface area to volume ratio and is permeable to respiratory gases. Gas exchange occurs across the entire cell membrane (Fig. 6.6).

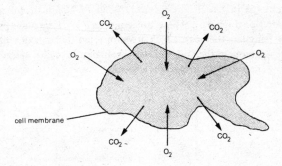

Fig. 6.6. *Gas exchange in* Amoeba

Similarly, in other simple aquatic organisms, e.g. *Hydra* (page 328), gas exchange occurs by diffusion over the entire body surface. Please note, these animals do not show breathing movements.

Gas exchange in the earthworm

In the earthworm, as in *Amoeba* and *Hydra*, the entire body surface is the gas exchange surface. The earthworm is much larger, however, and a transport system is needed to ensure that internal tissues receive enough oxygen. The skin epidermis is one cell thick and is kept moist by mucus secreted from gland cells (see Fig. 6.7). Oxygen dissolves in

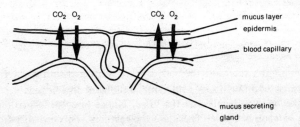

Fig. 6.7. *V.S. skin of earthworm showing gas exchange*
(NOTE. *Frogs have a similar arrangement*)

the surface moisture of the epidermis and diffuses into blood capillaries just below. Carbon dioxide diffuses out in the opposite direction. The blood system, containing dissolved haemoglobin, conveys oxygen to all parts of the body.

Gas exchange in insects

Gas exchange in insects occurs at the end of small air tubes called **tracheoles** (see Fig. 6.8). These are connected to larger air passages, **tracheae** (singular trachea), which are supported by rings of chitin. The tracheae open out at the body surface through pores called **spiracles**. The whole system is called the **tracheal system**.

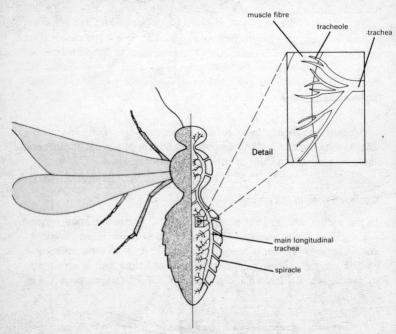

Fig. 6.8. *Tracheal system of an insect* (*right side only shown*)

To minimize water loss the body surface is covered with a waterproof cuticle and the spiracles are kept closed as much as possible.

In smaller insects, oxygen reaches the tracheole endings, from the atmosphere, by diffusion alone. Carbon dioxide travels in the reverse direction, from tracheoles, to tracheae and then through spiracles, again by diffusion. The tracheole ends are permeable to gases and are bathed in tissue fluid. Oxygen is supplied *directly* to body tissues, and *not* via a transport system. In insects, the blood does *not* transport respiratory gases.

Some insects show pumping movements of the abdomen which squeeze air in and out of the tracheal system. This ventilation (breathing) aids diffusion and increases the rate of gas exchange.

Gas exchange and breathing in fish

In fish, the site of gas exchange is the epithelium of gill filaments. It is moist, thin-layered, has a large surface area and is richly supplied with blood capillaries. It is also well ventilated.

On each side of the pharynx are four gills separated by five gill slits. The slits connect the pharynx to the water outside via the opercular valve (see Fig. 6.9). Each gill is made up of a large number of flat **gill filaments**, stacked close together and attached to the **gill arch**. Raised regions of the gill filaments, called **gill lamellae**, increase the surface area even further.

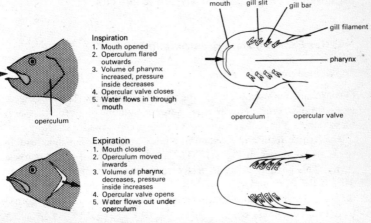

Inspiration
1. Mouth opened
2. Operculum flared outwards
3. Volume of pharynx increased, pressure inside decreases
4. Opercular valve closes
5. Water flows in through mouth

Expiration
1. Mouth closed
2. Operculum moved inwards
3. Volume of pharynx decreases, pressure inside increases
4. Opercular valve opens
5. Water flows out under operculum

Fig. 6.9. *Breathing in a bony fish*

A flow of oxygen-rich water over the gills is maintained by changes in volume of the pharynx which in turn generate pressure changes inside. During inspiration, the mouth is open and muscular action increases the volume of the pharynx, decreases the pressure inside it, and water flows in through the mouth. During expiration, the mouth is closed and muscular action decreases the volume of the pharynx, increases the pressure inside it, and water flows out through the open opercular valve.

The gas exchange surface, the gill epithelium, is only one cell thick. Oxygen dissolved in the water diffuses into the blood and combines with haemoglobin. The oxygen is carried in the blood to all parts of the body. Carbon dioxide in the blood diffuses out through the gas exchange surface and into the water. The water expelled under the operculum is rich in carbon dioxide.

Gas exchange in the frog

The adult frog has three respiratory surfaces: the lining of the mouth, the lungs and the skin. Gas exchange across the mouth lining is used when the frog is on land and restful. The lining is ventilated by visible throat movements. The lungs are small and simple and are used when the frog is active on land. The skin is the main respiratory surface and is used at all times, including swimming or hibernation. It has a thin epidermis covered with mucus, and supplied by a network of capillaries, as in the earthworm (page 119). A blood transport system containing haemoglobin conveys oxygen to all parts of the body.

6.5. *Gas Exchange and Breathing in Man*

The human respiratory system

Gas exchange in man occurs at the **alveoli** of the **lungs**. The lungs have an enormous surface area produced by the millions of alveoli (air sacs) they contain. The lungs are deep inside the body to reduce water loss.

The human respiratory system (the lungs and the air passages leading to them) are shown in Fig. 6.10. The lungs occupy most of the **thoracic cavity**. This airtight cavity is protected by the sternum and ribs (page 216) at the front and sides, and by the vertebral column (page 217) at the back. The muscular diaphragm separates the thoracic cavity (thorax) from the abdominal cavity.

Notice the lungs are separated from the ribs by a space, the **pleural cavity**, bordered on inside and outside by the **pleural membranes**. These membranes prevent the lungs being damaged by the ribs. The pleural cavity contains a fluid which lubricates the membranes so they move past each other easily.

On breathing in (inspiration) air reaches the alveoli by first entering

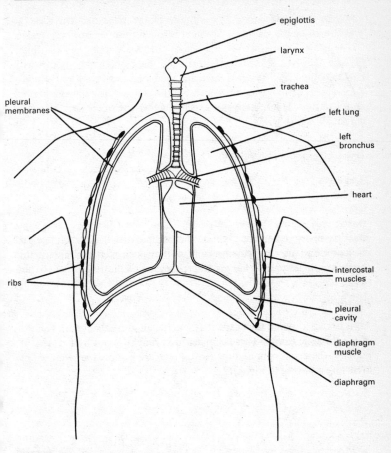

Fig. 6.10. *The human respiratory system*

the mouth or nostrils, passing through the **pharynx**, down the **trachea** (windpipe), and then through the two **bronchi** and into thousands of **bronchioles** in the lungs. The bronchioles end in groups of alveoli looking rather like microscopic bunches of grapes (see Fig. 6.11). Each alveolus is covered by capillaries and it is here that gas exchange takes place.

As air from the atmosphere enters the respiratory system it is warmed, and dust particles and bacteria are trapped by mucus secreted in the

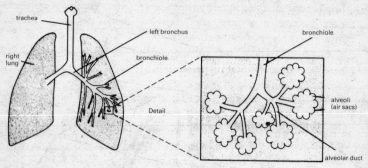

Fig. 6.11. *Detail of human lung showing smaller air passages and alveoli*

nasal passages and trachea. Cilia (microscopic hair-like structures but *not* hairs) line these air passages and as they beat, waft the mucus and trapped particles to the pharynx where they are swallowed.

Breathing

Air is drawn into and then expelled from the lungs by changes in volume of the thoracic cavity. These changes are brought about by two sets of muscles; the **diaphragm** and **intercostal muscles**. Follow the sequence of events by referring to Fig. 6.12.

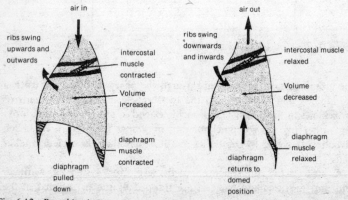

Fig. 6.12. *Breathing in man*
(NOTE. *Only 2 ribs are shown*)

Inspiration (breathing in): diaphragm and intercostal muscles contract, thorax increases in volume, pressure inside thorax decreases, and air moves into the lungs and they inflate.

Expiration (breathing out): diaphragm and intercostal muscles relax, thorax decreases in volume, pressure inside thorax increases, and air moves out of the lungs and they deflate.

It is essential to get these sequences right. There are two points to notice:

1. Inspiration is the active part of the breathing cycle, involving muscle contraction, whereas expiration is largely passive.

2. When volume increases, pressure decreases, and vice versa.

During normal breathing (what you are doing now) an adult man breathes in and out of the lungs about 0·5 litres of air in each breath. This is **tidal air**. During very deep breathing a maximum of 4 litres of air can be breathed in and out. This is called the **vital capacity**. One litre of air, the **residual air**, always remains in the lungs trapped by friction and cannot be breathed out. The **total capacity** of the lungs, the vital capacity plus residual air, is 5 litres (see Fig. 6.13).

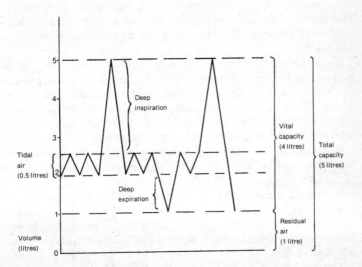

Fig. 6.13. *Volumes of air in the lungs*

You need to understand what these terms refer to, and what volumes they represent. The volumes given are approximate values for an adult man.

Control of breathing rate

The normal breathing rate is 15–20 breaths/minute but is higher during and just after periods of physical exercise. Breathing rate is controlled by the **respiratory centre** in the medulla of the brain. This centre is sensitive to carbon dioxide level in the blood. If the level increases, e.g. during vigorous exercise, nerve impulses from the brain to the intercostal and diaphragm muscles cause an increase in ventilation rate (rate of breathing). This in turn speeds up gas exchange to remove the excess carbon dioxide (and supply more oxygen). At the same time nerve impulses from the brain speed up heartbeat rate and the transport of respiratory gases.

As carbon dioxide levels return to normal so breathing rate and heartbeat return to normal. This control of carbon dioxide level in the blood is an example of **homeostasis** (see pages 164 and 202).

Gas exchange at alveoli

There are over 500 million alveoli in the human lungs. Alveoli have all the features we have come to expect of a respiratory surface in a large, active animal. They are thin, moist, have a large surface area, and are freely permeable to gases. In addition, they are well ventilated and have a rich blood supply; both are features which maintain a steep concentration gradient.

The air in the alveolus (alveolar air) has a higher concentration of oxygen than the arriving blood in the overlying capillary. Oxygen dissolves in the moisture lining the alveolus and diffuses across two layers of cells (the lining of the alveolus and the lining of the blood capillary) and into the blood. Here it diffuses into red blood cells (page 133) and combines with **haemoglobin** to form oxyhaemoglobin.

Haemoglobin is a respiratory pigment and because of its attraction for oxygen, enables the blood to carry large amounts of the gas. Haemoglobin gives blood its red colour.

Carbon dioxide is carried in the blood as the **bicarbonate ion**. Blood arriving at the alveolus is rich in carbon dioxide. The concentration of carbon dioxide in the alveolar air is lower. At the alveolus the bicarbonate breaks down to release carbon dioxide, which diffuses from the blood into the alveolus.

The blood, now rich in oxygen (oxygenated) and low in carbon dioxide is carried away to supply tissues throughout the body. The path taken by blood is shown on page 140.

The composition of alveolar air remains nearly constant (see Table

6.1) despite the continual loss of oxygen to the blood and gain of carbon dioxide. Replacement oxygen diffuses across from inspired air and carbon dioxide is lost to expired air (see Fig. 6.14).

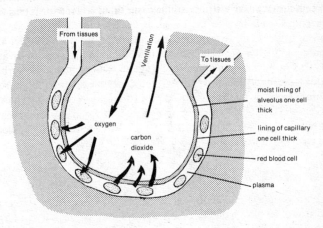

Fig. 6.14. *Gas exchange at an alveolus*

Table 6.1. *Composition of gases in breathed air*

	Inspired air	Expired air	Alveolar air
Oxygen	21%	16%	14%
Carbon dioxide	0.04%	4%	6%
Nitrogen	79%	79%	80%
Water vapour	Variable	Saturated	Saturated

You need to know the above table.

Now, using your knowledge of human breathing, gas exchange and gas transport, answer the following question.

Q. With the aid of diagrams, explain how, in a mammal, oxygen
(a) reaches the lungs
(b) enters the bloodstream, and
(c) is transported to the liver. [12, 8, 5]
[LON]

(In answering part (c) refer to page 140.)

Supplying tissues with oxygen and removing carbon dioxide

On arrival at the site of respiring tissues in the body, the oxyhaemoglobin breaks down to release oxygen (see Fig. 6.15). The oxygen diffuses through the capillary wall, across the tissue fluid, and into respiring cells.

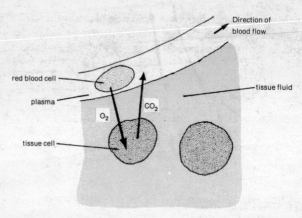

Fig. 6.15. *Delivery of oxygen and removal of carbon dioxide from tissues*

At the same time, carbon dioxide diffuses back in the other direction and, on reaching the blood, combines with water to form bicarbonate ions. The blood, now rich in carbon dioxide and low in oxygen (deoxygenated), travels through the circulatory system back to the lungs (see page 140).

Remember, oxygen is carried in the blood as oxyhaemoglobin within red blood cells. Haemoglobin picks up and releases oxygen under the following conditions:

$$\text{Haemoglobin + oxygen} \underset{\substack{\text{low } [O_2] - \text{ at body tissues} \\ \text{high } [CO_2]}}{\overset{\substack{\text{high } [O_2] - \text{ at the lungs} \\ \text{low } [CO_2]}}{\rightleftharpoons}} \text{Oxyhaemoglobin}$$

Carbon dioxide is carried in the blood as bicarbonate which is formed and broken down under the following conditions:

$$\text{Carbon dioxide + water} \underset{\substack{\text{high } [O_2] - \text{ at the lungs} \\ \text{low } [CO_2]}}{\overset{\substack{\text{low } [O_2] - \text{ at body tissues} \\ \text{high } [CO_2]}}{\rightleftharpoons}} \text{Bicarbonate}$$

The enzyme **carbonic anhydrase**, found in red blood cells, catalyses the above reaction.

6.6. *Gas Exchange in Flowering Plants*

The main gas exchange surface in flowering plants is the surface of the **spongy mesophyll cells** in the leaf. Gases enter from, and escape to, the atmosphere through the **stomatal pores** (see Fig. 6.16 and also 5.2, page 66).

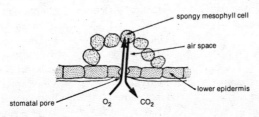

Fig. 6.16. *Gas exchange at a spongy mesophyll cell during darkness*

Gas exchange in green plants is complicated by the process of photosynthesis, which in many ways is the reverse of respiration. The overall exchange of gases varies with light intensity (see page 79).

During daylight hours, the rate of photosynthesis is usually much greater than respiration, so overall the mesophyll cells take up carbon dioxide and give off oxygen. This is the reverse of the direction of gas movement at the respiratory surface of animals.

During darkness, plants respire but do not photosynthesize, and so there is an overall release of carbon dioxide and uptake of oxygen. This is the same as that at the respiratory surface of animals.

In woody plants (page 320) gas exchange also occurs at the **lenticels** of the stem (Fig. 6.17). These small openings in the bark of woody stems are filled with loosely packed cells separated by air spaces.

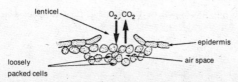

Fig. 6.17. *Gas exchange at a lenticel in the stem of a woody plant*

Definitions

Respiration
Gas exchange
Breathing (ventilation)

Key Words

Respiratory or
  gas exchange surface
Mitochondria
ATP
Anaerobic
Aerobic
Alcohol (ethanol)
Lactic acid
Lime water
Soda lime
Potassium hydroxide
Haemoglobin
Epidermis
Tracheoles
Tracheae

Spiracles
Tracheal system
Gill
Gill filaments
Gill arch
Gill lamellae
Pharynx
Operculum
Alveoli
Lungs
Thoracic cavity
Pleural cavity
Pleural membranes
Trachea (windpipe)
Bronchi

Bronchioles
Diaphragm
Intercostal muscles
Inspiration
Expiration
Tidal air
Vital capacity
Residual air
Total capacity
Bicarbonate
Carbonic anhydrase
Spongy mesophyll
  cells
Stomatal pores
Lenticels

Exam Questions

1. (a) Give a balanced equation which summarizes the process of tissue respiration.   [5]
(b) What are the differences between aerobic and anaerobic respiration?
                                                                        [7]
(c) Describe an experiment which demonstrates that germinating seeds give off heat.   [9]
(d) How does a named protozoan obtain its oxygen supply?   [4]
                                                                    [LON]

2. (*a*) *Name* precisely *the gas exchange surface for the following animals*
(*i*) *a mammal*   [1]
(*ii*) *a fish*   [1]
(*iii*) *an insect*   [1]
(*iv*) *Name* three *features shared by* all *the above animals' gas exchange surfaces and* one *feature shared by* only two *of them* (*state which two share this feature*).   [5]

[AEB]

3. *Explain how* (*i*) *a filamentous alga*
(*ii*) *an adult amphibian*
(*iii*) *a flowering plant and*
(*iv*) *a mammal, obtain oxygen*   [4, 3, 6, 8]

[LON]

(See page 312 for part (i).)

7. Transport

Every living cell in an organism needs to take up substances, such as food and oxygen, and has to get rid of waste. These substances enter and leave cells mainly by diffusion (page 55).

In single-celled organisms, such as *Amoeba* (page 326), *Chlamydomonas* (page 310) and *Spirogyra* (page 312), diffusion through the body surface and across the cell membrane supplies the needs of the organism. This also applies to the simpler multicellular organisms such as *Hydra* (page 328) and *Taenia* (page 330), which have a thin body wall.

In larger multicellular organisms, however, cells on the inside are too far away from the external environment to obtain supplies and get rid of wastes simply by diffusion. Thus, special transport systems have developed to deliver and take away substances. A transport system links the cells inside the body to the external environment via specialized structures. In mammals, for example, the lungs exchange respiratory gases, the gut absorbs soluble food, and the kidneys remove wastes. In higher plants, for example, root hairs absorb water and minerals from the soil. Such structures perform their functions on behalf of all cells in the body.

In certain invertebrate groups, e.g. the annelids (page 304) and arthropods (page 305), and in all vertebrates (page 305), the transport system consists of a fluid (blood) pumped through vessels by a muscular pump (the heart). This is called a **circulatory system**.

In flowering plants (page 319), there are two transporting fluids; the phloem and xylem. Both flow through systems of tubes (see page 149).

A detailed knowledge and understanding of the transport systems of both mammals and higher plants is required.

The components of the mammalian circulatory system are:
1. circulatory fluids (blood and lymph)
2. a system of vessels to carry the fluids (arteries, veins, capillaries and lymph vessels)
3. a pump (the heart).

7.1. *The Mammalian Circulatory Fluid: Blood*

Components of blood

Blood is a tissue (page 41) made up of three parts:
1. a liquid part, **plasma**, making up 55 per cent of blood
2. **blood cells** of several types, making up 45 per cent of blood
3. small cell fragments called **platelets.**

1. **Plasma** is a straw-coloured solution, 90 per cent water, and contains the following dissolved substances:
(a) mineral salts, e.g. sodium chloride ($NaCl$) and sodium bicarbonate ($NaHCO_3$)
(b) food substances – monosaccharides, glucose, amino acids, fatty acids, glycerol and vitamins
(c) plasma proteins – fibrinogen and antibodies
(d) hormones (chemical messengers)
(e) nitrogenous wastes, mainly urea.

The relative amounts of these substances vary from one part of the body to another (page 143) and from one minute to the next. For example, the glucose level in plasma increases after a meal and the level of the hormone adrenalin is increased by emotional stress (page 200).

2. **Blood cells**: there are two main types:
(a) red blood cells (erythrocytes)
(b) white blood cells (leucocytes).

The features of these cells are summarized in Table 7.1.

Red blood cells are minute biconcave discs. They have no nucleus and hence have a short life span (about 3 months). They are broken down by the liver (page 96) and are constantly replaced by cells in the bone marrow. Red blood cells contain two important substances: **haemoglobin**, which combines with oxygen for transport (page 126), and **carbonic anhydrase**, which converts carbon dioxide to bicarbonate for transport (page 129).

White blood cells are larger than red blood cells, have a nucleus and are present in much fewer numbers. There are three main types and each plays a vital role in the body's defence against disease (page 147).

3. **Platelets** (thrombocytes): these are very small cell fragments formed by cells in the bone marrow. At the site of an injury, platelets, together with damaged cells, release an enzyme **thrombokinase** which initiates blood clotting. A sequence of events follows:

Table 7.1. The structures in mammalian blood

	Appearance	Number/mm³	Site of Formation	Function
Red blood cells (erythrocytes)	i) biconcave (thin central region) ii) no nucleus iii) red due to presence of haemoglobin	5 million	Red bone marrow	Oxygen transport (see text) Bicarbonate formation (see text)
White blood cells (leucocytes) Two main types:				
1. Granulocytes	i) granular (grainy) cytoplasm ii) lobed (lumpy) nucleus	7,000	Red bone marrow	Phagocytic — engulf bacteria
2. Agranulocytes Two types: a) Lymphocytes	i) non-granular cytoplasm ii) large nucleus	2,500	Lymph nodes	Secrete antibodies
b) Monocytes	i) non-granular cytoplasm ii) kidney-shaped nucleus	500	Red bone marrow	Phagocytic — engulf bacteria
Platelets	i) cell fragments	¼ million	Red bone marrow	Blood clotting

Note: In a sample of normal human blood the above structures are present in the ratio red blood cell : white blood cell : platelet
500 : 1 : 25

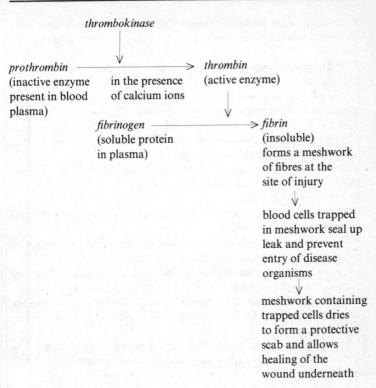

thrombokinase

prothrombin ————————————→ thrombin
(inactive enzyme   in the presence   (active enzyme)
present in blood   of calcium ions
plasma)

fibrinogen ————————————→ fibrin
(soluble protein   (insoluble)
in plasma)   forms a meshwork
                                           of fibres at the
                                           site of injury

                                           blood cells trapped
                                           in meshwork seal up
                                           leak and prevent
                                           entry of disease
                                           organisms

                                           meshwork containing
                                           trapped cells dries
                                           to form a protective
                                           scab and allows
                                           healing of the
                                           wound underneath

This sequence of events is complex, but is necessary to make sure clotting does not take place by accident within blood vessels. A clot blocking the inside of a blood vessel stops the flow of blood. If this occurs in the brain it can cause a 'stroke', or in the heart muscle, a coronary thrombosis (heart attack).

Functions of blood

1. **Formation of tissue fluid**: every cell in the body of a mammal lives in a liquid environment called tissue fluid, which is derived from blood. Tissue fluid is basically blood plasma that has been forced out of blood capillaries (see page 144).

Blood has several major transport functions:

2. It **transports oxygen** from the lungs to respiring tissues in the form of oxyhaemoglobin within red blood cells (page 126).

3. It **transports carbon dioxide** from respiring tissues to the lungs in the form of bicarbonate (page 126).

4. It **transports the products of digestion** (glucose, amino acids, fatty acids, glycerol) from the small intestine to the liver, and then to the rest of the body.

5. It **transports nitrogenous wastes** (e.g. urea from the liver) from tissues to the kidneys.

6. It **transports hormones** from the glands which secrete them (endocrine glands) to the target tissues on which they act.

Blood distributes heat and is involved in regulating body temperature:

7. The heat generated by active organs such as the liver and muscles is distributed around the body by the blood.

8. By varying the amount of blood which flows to the skin, heat loss from the body can be controlled. In this way, body temperature can be regulated (page 176).

The blood plays a major role in protecting the body against pathogens (disease-causing organisms):

9. Two types of white blood cell, granulocytes and monocytes (page 134) engulf pathogens by phagocytosis.

10. Another type of white blood cell, the lymphocytes (page 134), secrete chemicals called antibodies which attack pathogens (page 147).

11. The plasma protein fibrinogen initiates blood clotting at the site of an injury (page 135). Clotting helps prevent invasion of the body tissues by pathogens.

Notice 'Functions of blood' has eleven points grouped under four sections. With proper explanation and attention to detail you should be able to give clear and full answers to questions such as:

Q. Explain how the circulatory system of a mammal performs its functions of

(a) *transport*

(b) *defence.*   [17, 8]

[LON]

(Refer also to pages 137 and 147 when answering this question.)

7.2. *Blood Vessels*

Muscular contractions of the heart pump blood into **arteries**, which divide inside body tissues and organs to form first arterioles, and then very fine vessels called **capillaries**. These join together to form first venules and then **veins**, which carry the blood back to the heart.

At the arterial end of capillaries, plasma is forced into the surrounding tissues to form **tissue fluid**. Most of this fluid returns to the blood at the venous end of capillaries, but some enters **lymph vessels** to form **lymph** (page 145). The lymph travels through the lymphatic system and empties into the bloodstream just before the right side of the heart.

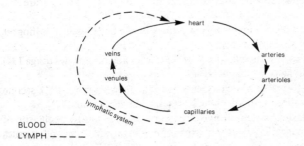

BLOOD ———
LYMPH — — — —

The structure and functions of the different types of blood vessel are summarized in Table 7.2. Notice how structure is related to function.

Arteries have thick walls to withstand high blood pressure and their elastic walls convert the irregular blood flow from the heart into a steady flow to the tissues. Elastic recoil of the arteries gives rise to the pulse which can be felt in the neck and wrist.

The capillaries are small in diameter and very 'leaky'. All the blood's functions depend on the permeability of capillaries, since it is here that substances are exchanged with surrounding tissues.

Blood in veins is at low pressure and its flow towards the heart is assisted by one-way valves and by the action of contracting skeletal muscles (page 221) which squeeze on the veins.

Mammals have a **double circulatory system** consisting of two sets of blood vessels connected to the heart. The **pulmonary circulation** carries blood to and from the lungs, while the **systemic circulation** conveys blood to and from tissues throughout the body (see Fig. 7.1).

Having two circulations keeps oxygenated blood separate from de-oxygenated blood. Also, it ensures blood is delivered to tissues at high pressure. After passing through one set of capillaries the blood is returned to the heart to have its pressure raised.

Table 7.2. Mammalian blood vessels: their structure and function

	Structure	Function
Arteries 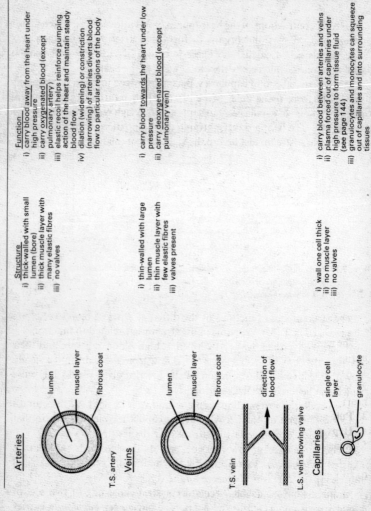 T.S. artery	i) thick-walled with small lumen (bore) ii) thick muscle layer with many elastic fibres iii) no valves	i) carry blood away from the heart under high pressure ii) carry oxygenated blood (except pulmonary artery) iii) elastic recoil helps reinforce pumping action of the heart and maintain steady blood flow iv) dilation (widening) or constriction (narrowing) of arteries diverts blood flow to particular regions of the body
Veins T.S. vein L.S. vein showing valve	i) thin-walled with large lumen ii) thin muscle layer with few elastic fibres iii) valves present	i) carry blood towards the heart under low pressure ii) carry deoxygenated blood (except pulmonary vein)
Capillaries	i) wall one cell thick ii) no muscle layer iii) no valves	i) carry blood between arteries and veins ii) plasma forced out of capillaries under high pressure to form tissue fluid (see page 144) iii) granulocytes and monocytes can squeeze out of capillaries and into surrounding tissues

Labels (Arteries T.S.): lumen, muscle layer, fibrous coat

Labels (Veins T.S.): lumen, muscle layer, fibrous coat

Labels (L.S. vein showing valve): direction of blood flow

Labels (Capillaries): single cell layer, granulocyte

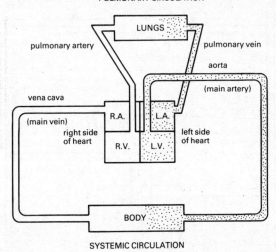

Fig. 7.1. *The double circulation of mammals (oxygenated blood shown stippled)*
  R.A. = Right atrium   *L.A. = Left atrium*
  R.V. = Right ventricle   *L.V. = Left ventricle*

The major blood vessels and their connections with the heart are shown in Fig. 7.2. The following points should help you remember their names:
1. **Arteries** carry blood away from the heart and to organs.
2. **Veins** (except portal veins) carry blood away from organs and to the heart.
3. **Pulmonary** refers to the lungs.
4. **Hepatic** refers to the liver.
5. **Renal** refers to the kidneys.
  Questions may be asked about the route blood takes from one organ to another. For example:

Q. What is the correct sequence of blood vessels through which blood travels from the left ventricle to the kidneys and back to the heart in a mammal?
(a) *Dorsal aorta, renal artery, renal vein, posterior vena cava.*
(b) *Dorsal aorta, hepatic artery, hepatic portal vein, posterior vena cava.*
(c) *Anterior vena cava, renal vein, hepatic artery, dorsal aorta.*
(d) *Anterior vena cava, hepatic vein, renal artery, pulmonary vein.*
(e) *Hepatic portal vein, hepatic vein, dorsal aorta, renal artery.*   [1]

[LON]

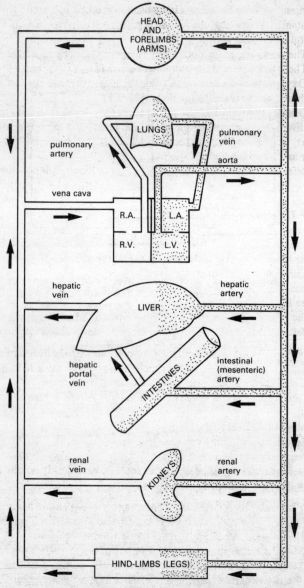

Fig. 7.2. *Connections between the heart and main blood vessels in man*
(arrows show direction of blood flow; oxygenated blood shown stippled)

Questions may be asked about the route taken by certain dissolved substances. In this case, the following points will help you (refer to Fig. 7.2 as you go through them):

1. The **veins** (except the pulmonary vein) carry deoxygenated blood.
2. The **arteries** (except the pulmonary artery) carry oxygenated blood.
3. **Oxygen** takes a short route from the lungs, through the left side of the heart, before reaching the body tissues.
4. **Dissolved foods** take a long route from the small intestine, through the liver, the right side of the heart, the lungs, and the left side of the heart, before reaching the body tissues (most fats, however, bypass the liver and travel through the lymphatic system).

Refer also to section 7.4 below, 'Changes in Blood Composition'.

7.3. *The Heart*

The mammalian heart is a muscular pump situated between the lungs in the thoracic cavity. Its function is to:

1. circulate blood through the blood vessels
2. generate a blood pressure sufficient to force plasma out of capillaries (page 144).

The mammalian heart is actually two pumps joined together, one on the left and one on the right. Both pumps have two chambers, an **atrium** (auricle) and a **ventricle**. Valves in the heart ensure that blood flows through the heart in one direction only.

An accurate, fully labelled drawing of the heart should be learnt – you may be expected to draw or label one. Before doing this, it is worth looking at a simplified diagram of the heart showing the basic features (see Fig. 7.3 overleaf).

Compare this figure with the one below it showing the mammalian heart in ventral view.

Notice the following features:

1. The right ventricle pumps blood to the lungs.
2. The left ventricle pumps blood round the body.
3. The atria have thinner walls than the ventricles.
4. The left ventricle has a thicker muscular wall than the right ventricle, because it must generate a force sufficient to pump blood round the entire body.
5. One-way valves are positioned at the exits of the heart chambers.

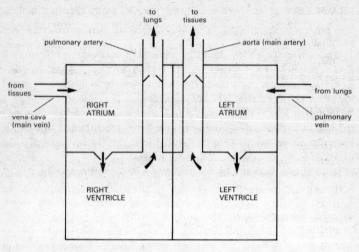

Fig. 7.3. *The mammalian heart (diagrammatic): ventral (front) view*

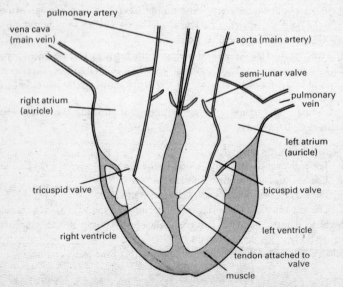

Fig. 7.4. *Vertical section through the mammalian heart: ventral (front) view*

Cardiac cycle

The pumping action of the heart is described by the cardiac cycle. The cycle lasts just under one second and consists of a contraction phase, **systole**, followed by a relaxation phase, **diastole**. Each phase lasts about half a second.

Control of heartbeat rate

The average heartbeat rate in an adult human is about 70 beats/minute, but changes according to the body tissues' requirement for food and oxygen, and their need to remove carbon dioxide.

Heart muscle, unlike other muscle types (page 221), is **myogenic**, that is, it contracts by itself without nervous stimulation. Nevertheless, the heart is connected to the autonomic nervous system (page 187), which regulates its rate of beating. Hormones also affect the heartbeat rate.

At times of stress or danger, the adrenal glands (page 199) release the hormone adrenalin. This stimulates the heart to increase its rate of beating to prepare the body for 'fight or flight' (page 200).

When the level of carbon dioxide in the blood rises, for example during exercise, then the heart is stimulated by the autonomic nerve supply to increase its rate of beating.

7.4. *Changes in Blood Composition*

As blood passes round the body, its composition changes depending on the function(s) of the organ or tissue through which it is flowing. For example, when blood flows through any tissue it loses oxygen and dissolved foods, and gains carbon dioxide. The overall changes in blood composition can be summarized as follows:

Region of the body	Blood gains	Blood loses
All tissues	Carbon dioxide Nitrogenous waste	Oxygen Food (monosaccharides, amino acids, etc.)
Lungs	Oxygen	Carbon dioxide Water
Small intestine	Dissolved foods – water, minerals, vitamins, monosaccharides, amino acids, glycerol, fatty acids	

Region of the body	Blood gains	Blood loses
Liver	Controlled amounts of glucose and fats Urea Heat	Glucose (for storage as glycogen) Fatty acids and glycerol Excess amino acids Old red blood cells
Kidneys		Urea Water Salts
Bones	Red blood cells Granulocytes and monocytes (types of white blood cell)	Iron (for Hb in red blood cells) Calcium and phosphate (for bone growth)
Skin	Vitamin D	Heat Salts and urea (in sweat)
At the veins of the neck where the lymphatic system empties into the bloodstream	Lymph containing fats and lymphocytes	

This list is by no means complete but does show the major changes. Hormones (page 198) have not been included.

7.5. *Formation of Tissue Fluid*

All the cells in a mammal's body are no more than 0·025mm from a capillary – the source of the tissue fluid which surrounds them. All the cells' food and oxygen are obtained from the tissue fluid, and all the cells' wastes are lost into it.

Tissue fluid is formed at the arterial end of capillaries where the high blood pressure forces fluid out through the thin walls of the capillary and into spaces between the cells (Fig. 7.5). It is similar to plasma in containing oxygen and dissolved foods but does not contain protein molecules because they are too large to pass through capillary walls. Red blood cells, platelets and the non-phagocytic white blood cells, lymphocytes, are similarly unable to pass through the capillary walls.

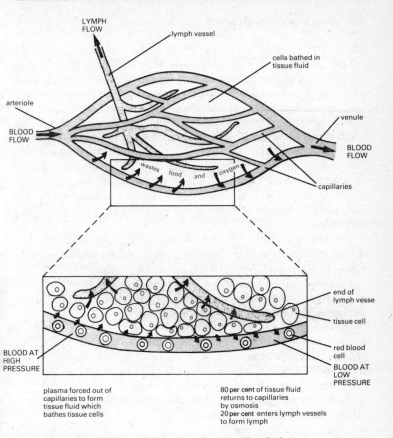

Fig. 7.5. *Formation of tissue fluid and lymph at a capillary bed*

Phagocytic white blood cells, the granulocytes and monocytes (page 134), can squeeze out of capillaries and into tissues.

At the venous end of capillaries the blood pressure is much reduced, and the blood (now very thick because of the plasma it has lost) absorbs back tissue fluid by osmosis (see page 57). Some of the tissue fluid filters into small blind-ended lymph vessels to become **lymph**.

7.6. *The Lymphatic System*

The lymphatic system is a very extensive system of vessels which drains off tissue fluid from capillary beds all over the body. The lymph so formed is similar in composition to tissue fluid but in addition contains lymphocytes and certain proteins. Where the lacteals (page 91) empty into the lymphatic system, the lymph contains a large amount of fat.

At intervals along the lymph vessels are collections of cells forming **lymph nodes**. These manufacture lymphocytes and antibodies (page 148) and contain phagocytic cells which filter out, ingest and destroy bacteria. Lymph nodes in the armpits, groin and neck are grouped together to form lymph glands. The tonsils and spleen are essentially large lymph glands. Lymph nodes near the site of an infection often become swollen and painful as a result of the increased activity of the white blood cells they contain.

The flow of lymph in lymph vessels is maintained by the action of skeletal muscles pressing on the vessels and by the presence of one-way valves. In this respect, lymph vessels are similar to veins (page 138).

The lymph vessels eventually join to form two large lymphatic ducts which return lymph to the bloodstream at the major veins of the neck just before the vena cava.

The main functions of the lymphatic system are:

1. To return to the bloodstream substances which do not re-enter the venous end of capillaries.

2. To absorb fats from lacteals in the villi of the small intestine and then transport and empty them into the bloodstream.

3. To destroy bacteria by producing white blood cells which:

(a) filter out and engulf bacteria

(b) secrete antibodies (page 148).

Finally, remember lymph is derived from tissue fluid which in turn is derived from blood plasma. Comparisons between plasma, tissue fluid and lymph are popular exam questions.

	Blood plasma	*Tissue fluid*	*Lymph*
Location	Heart and blood vessels	Tissues	Lymphatic system
Differences in contents	Plasma proteins Fat droplets Red blood cells	No plasma proteins No fat droplets No red blood cells	Some proteins Fat droplets No red blood cells

7.7. *Defence Against Disease*

The mammalian body is under continual bombardment by microscopic organisms which can cause disease if they enter the body and multiply. The four commonest types of disease-causing organism (**pathogen**) are:
1. certain bacteria (page 307)
2. viruses (page 305)
3. certain protozoa (page 304)
4. certain fungi (page 304).

The majority of infectious human diseases are caused by bacteria or viruses.

To combat these organisms the body has a sophisticated system of defence. The first line of defence attempts to prevent organisms entering the body, and the second attempts to kill or immobilize them if they do enter.

First line of defence

Physical barriers and active agents normally prevent pathogens from entering the body.

Physical barriers:
1. The skin epidermis (page 174) is impermeable.
2. Healthy mucous membranes in the mouth, nose, windpipe and oesophagus are impermeable.
3. Hairs lining the nasal cavity filter out airborne particles and pathogens.

Active agents:
1. On the skin, the oily substance sebum produced by sebaceous glands (page 173) contains an anti-bacterial agent.
2. On the skin, sweat produced by sweat glands (page 173) contains an anti-fungal agent.
3. In the eyes, tears contain an anti-bacterial agent.
4. In the stomach, the acid-containing gastric juice kills bacteria.
5. In the nose and windpipe, cilia (page 124) trap airborne particles and waft them towards the mouth where they are swallowed.
6. If there is surface damage to the skin, blood clotting helps prevent entry of pathogens.

Second line of defence

If the first line of defence fails, and pathogens enter the body, the second line of defence is called into play:

1. Phagocytic cells in the lymph nodes, and in the blood (granulocytes and monocytes), engulf and destroy bacteria.
2. Interferon, a protein secreted by individual cells throughout the body, can stop the multiplication of viruses.
3. The immune system (see below) produces antibodies which attack pathogens.

Immunity

Definition: Immunity *is the body's ability to resist infection by the use of antibodies.*

Immunity to a particular disease may be acquired naturally or may be induced artificially.

The basis of immunity lies in the body's ability to manufacture **antibodies** which can attack **antigens**. Antigens are chemicals on the surface of, or secreted by, pathogens. Antibodies are made and secreted by lymphocytes (page 134).

Immunity can be classified as active or passive.

Active immunity is when an organism achieves immunity to a particular disease by making its own antibodies to combat the disease organism.

Passive immunity is when an organism gains immunity by receiving antibodies 'ready-made' from another organism.

Active immunity can be acquired naturally, by recovering from the disease, or induced artificially, by vaccination.

Examples of naturally acquired active immunity are resistance to measles, mumps or chicken pox, after recovering from the illness.

Alternatively, active immunity can be induced artificially by vaccination. **Vaccination** is the injection into the body of disease antigens which trigger the body's immune system to produce appropriate antibodies. Antigens of the disease organism can be given in the form of:

(a) Toxoids, which are toxins (poisons) produced by the pathogen and then inactivated (made harmless) before injection. The diphtheria vaccine is an example.

(b) Killed vaccines, which contain the dead pathogen. The poliomyelitis (polio) vaccine is an example.

(c) Attenuated (weakened) living vaccines, which contain a weakened strain of the pathogen. An example is the BCG vaccine against tuberculosis (TB).

Passive immunity can occur naturally or be induced artificially.

An example of naturally acquired passive immunity is the protection against infection a foetus gains from antibodies which are supplied by the mother (page 254).

A tetanus injection is an example of artificially induced passive immunity. After a surgical operation a person may be given an anti-tetanus vaccination. The vaccine contains blood serum from a horse which has been infected with the tetanus bacterium. (NOTE. Serum is plasma with the fibrinogen removed, so that it does not clot.) The horse has manufactured the appropriate anti-tetanus antibodies, and these are being given ready-made to the patient. The vaccine gives the patient short-term protection against tetanus infection.

7.8. *Transport in Plants: Xylem and Phloem*

In flowering plants (page 319) there are two transport systems: **xylem** and **phloem**. Xylem is the fluid found inside xylem vessels. It transport water and minerals from the roots to the stem and the leaves. Phloem in phloem sieve tubes transports food substances from the sites where they are made,

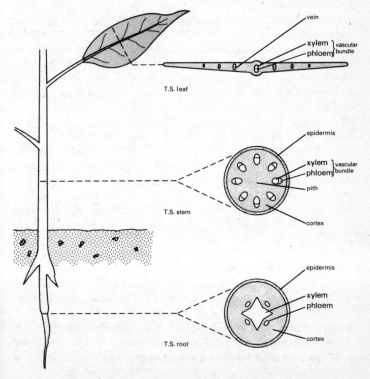

Fig. 7.6. *Distribution of phloem and xylem in a herbaceous dicot* (*page 319*)

notably the leaves, to the growing parts and storage regions of the plant.

Together, the xylem vessels and phloem sieve tubes form the tube-like **vascular bundles** which run through roots and stems and can be seen as veins in the leaf (Fig. 7.6). Each vascular bundle contains a number of xylem vessels and phloem sieve tubes (Fig. 7.7). In the root and stem of a dicot, the sieve tubes are to the outside of the xylem vessels (Fig. 7.6 and 7.7).

Xylem vessels are formed from once-living cells which have deposited **lignin** in their cell walls and have subsequently died. The lignin water-proofs and strengthens the cell walls and plays an important role in supporting the plant (page 229). After cell death, the cross walls break down (Fig. 7.7) and the string of cells running from root to leaf forms a continuous tube. At intervals, holes or pits enable substances to be transported sideways out of the vessel (page 159).

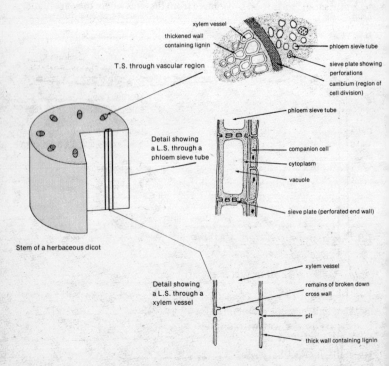

Fig. 7.7. *Detail of vascular tissues in a herbaceous dicot*

The phloem sieve tubes are made up of living cells. One cell is con-nected to the next by a perforated end wall called a sieve plate (Fig. 7.7). The cells have no nucleus and their cell walls do not contain lignin. Companion cells are closely associated with the sieve tube cells and are thought to be actively involved in the transport of substances within the tube.

The differences between xylem vessels and phloem sieve tubes can be summarized in a table:

	Xylem vessels	*Phloem sieve tubes*
Substances transported	Water and mineral salts	Organic substances, e.g. sucrose and amino acids
From	Roots	Sites of photosynthesis
Direction	Upwards	Up or down
To	Stem, leaves and associated structures	All parts of the plant, especially growing regions, e.g. root and shoot tips; storage regions, e.g. bulbs or tubers; and fruits and seeds
Cell structure	Cells non-living No cross walls	Cells living Perforated cross walls (sieve plates)
	Cell walls strengthened with lignin	No lignin

7.9. *Transport in Xylem Vessels*

Uptake and transport of water

The movement of water from the soil into the root xylem and up into the stem and leaves is maintained by four processes:
1. osmotic flow of water from soil to root cortex
2. root pressure
3. capillarity (capillary rise)
4. transpiration.

1. **Osmotic flow** from soil to root cortex (steps 1 and 2 in Fig. 7.8). Water is taken up (absorbed) from the soil and into the plant root by osmosis.

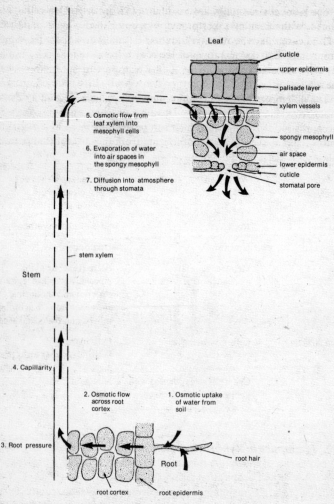

Fig. 7.8. *Summary of water uptake and transport in a flowering plant*

This occurs mainly through the root hairs of epidermal cells. Root hairs are found in a narrow zone 0·5 to 1 cm behind the root tip (page 323). They increase the surface area of the root for absorption of water, and several million are found in a single plant.

Soil water is usually found as a thin film around individual soil particles.

Root hair cells normally have a higher O.P. (osmotic potential) than the soil water and thus take up water by osmosis. Cells towards the centre of the cortex have a relatively high O.P. and water flows by osmosis from one cell to the next towards the root xylem.

2. **Root pressure** (step 3 in Fig. 7.8): it is not entirely clear how water passes from the root cortex into xylem vessels. It is believed an active transport process (page 61) pumps mineral ions into the root xylem and water then follows by osmosis. This flow of water into the root xylem accounts for root pressure, which can be demonstrated in a plant stem cut 2–3 cm above soil level. Sap oozing from the cut end is able to displace a mercury or water column (Fig. 7.9).

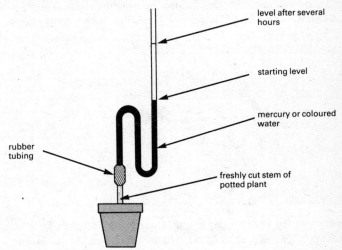

level after several hours

starting level

mercury or coloured water

rubber tubing

freshly cut stem of potted plant

Fig. 7.9. *Demonstration of root pressure*

Root pressure is particularly evident in deciduous trees in early spring. It causes sap to rise up the stem at a time when the tree is without leaves and little or no transpiration occurs (see over).

Root pressure is maintained by an active process which requires energy from respiration. If the root is deprived of oxygen or is treated with a respiratory inhibitor, root pressure disappears.

3. **Capillarity** (step 4 in Fig. 7.8) is the force of attraction between molecules in a liquid and the sides of a narrow tube, which results in the liquid rising up inside the tube. You have probably noticed this effect with drink rising up the inside of a drinking straw. The narrower the tube the higher the liquid will rise. This can easily be demonstrated using glass capillary tubes of different bore.

Capillarity certainly aids the rise of xylem up the xylem vessels of a plant stem, but the effect is small. Transpiration plays a much greater role.

4. **Transpiration** (steps 5–7 in Fig. 7.8) is the loss of water from the plant by evaporation. This evaporation occurs mainly from the surface of spongy mesophyll cells in the leaves, the water being lost mainly through stomata (page 129). A small amount of water is lost through the waxy cuticle of the leaf and through the cuticle and lenticels (page 129) of the stem.

Definition: Transpiration *is the evaporation of water from the aerial (above-ground) surfaces of the plant.*

Transpiration can be demonstrated by clamping cobalt chloride paper on to a leaf (Fig. 7.10). The paper is blue when dry but turns pink when

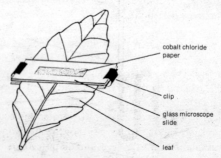

Fig. 7.10. *Using cobalt chloride paper to demonstrate water loss from a leaf*

moist. On the leaf, the paper usually turns pink after an hour or so, indicating that water is being lost from the leaf surface. If paper is clamped to both upper and lower surfaces, which paper would you expect to turn pink first? Why? (see page 65).

During transpiration, the spongy mesophyll cells lose water by evaporation and their O.P. rises. Water is drawn by osmosis from adjacent cells and then in turn from xylem vessels in the leaf. This creates a tension or pull on the xylem in the leaf and stem. This pull is sufficient to draw water up the stem in a continuous flow, called the **transpiration stream**.

The flow of water up the xylem vessels can be demonstrated by placing the cut end of a leafy shoot in water stained with a dye such as eosin red or methylene blue. After several hours, transverse sections of the stem are cut and examined under a microscope. It is seen that only xylem tissue contains the dye, indicating that it is xylem which conducts water up the stem.

Transpiration is a loss of valuable water, and if the water lost by evaporation is not replaced by water uptake, then the plant will wilt. Evaporation from a plant is an unfortunate side effect of having leaves which are designed to allow gas exchange to occur. A leaf structure which encourages gas exchange also allows rapid water loss (see page 129). However, transpiration may have certain useful effects. In hot conditions, it cools the leaf slightly (see also sweating in mammals on page 176). Also, it is argued, by producing the transpiration stream, it draws mineral salts up the plant.

Rates of transpiration

The rate of transpiration in a plant depends on the water diffusion gradient (page 56) between the air spaces in the leaf, and the atmosphere just outside the stomatal pore. Any factor which increases this gradient will increase transpiration.

The main factors governing transpiration rate are:
1. Temperature: a rise in temperature increases the rate of transpiration by speeding up the evaporation and diffusion of water.
2. Air humidity: if the air is moist (humidity high) the water diffusion gradient will be low and the transpiration rate low. The opposite applies if the air is dry. Dry air favours transpiration.
3. Air movement (wind) will carry away water molecules that have diffused through stomata. Strong air movements (wind) stop water molecules accumulating outside the stomatal pore. This helps maintain a steep diffusion gradient and thus increases transpiration.
4. Any factor which causes stomata to close will effectively stop transpiration. Light and water availability control opening and closure of stomata:
(a) In general, stomata open during daytime and close at night.
(b) Stomata close at times of severe water shortage, when wilting occurs.

In summary, transpiration rates are increased by a rise in temperature, a lowering in air humidity, or an increase in air movement. Transpiration rates are decreased by a drop in temperature, a rise in humidity, or a decrease in air movement. In addition, absence of light or a shortage of water causes stomata to close and thus stops transpiration.

The rate of transpiration under different conditions can be measured using a simple apparatus such as that shown in Fig. 7.11. A well-watered potted plant is used and the pot and soil enclosed in a waterproof plastic bag to reduce evaporation from the soil. The plant is weighed, left under experimental conditions for a set number of hours, and then reweighed. Any weight loss is presumed to be due to loss of water by evaporation from the stem and leaves, i.e. transpiration.

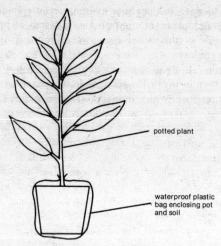

potted plant

waterproof plastic
bag enclosing pot
and soil

Fig. 7.11. *Apparatus for measuring water loss in a plant*

The experiment can be repeated under different sets of conditions. For example, the plant can be placed in normal light, and then darkness, and the transpiration rates compared. By placing the plant in a transparent enclosure, and introducing varying amounts of water, the plant can be exposed to different humidities. A hair drier (with the heating element switched off) can be used to generate varying wind conditions. In each case, the experiment should be repeated under the same laboratory conditions, with only one factor altered – light, temperature, humidity, or wind.

Typical results under laboratory conditions are:

Condition	Rate of transpiration as measured by weight loss
Normal light	High
Darkness	Low
High humidity	Low
Low humidity	High
High wind	High
Low wind	Low
High temperature	High
Low temperature	Low

The amount of water lost per hour by transpiration is small compared to the overall weight of plant and pot. Using the above apparatus, experiments should be carried out over several hours or days to get a measurable weight loss.

Rates of transpiration can be measured more rapidly using a **potometer** such as that shown in Fig. 7.12. This apparatus measures water uptake rather than water loss. Water uptake and loss are not exactly equal, i.e. not all the water taken up by the plant is lost by evaporation. Some water, but only about 5 per cent, is used by the plant in photosynthesis and in forming new tissue. Transpiration rates measured using a potometer are very slightly overestimated, but the difference is negligible.

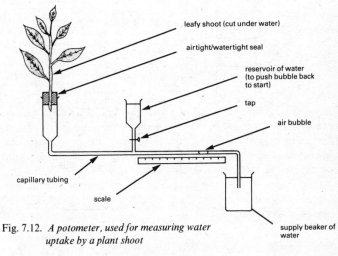

Fig. 7.12. *A potometer, used for measuring water uptake by a plant shoot*

Certain precautions are necessary when using a potometer. Refer to Fig. 7.12 as you go through them.
1. The stem of a leafy shoot is used. It must be cut under water and kept immersed while being connected to the apparatus. This prevents air from entering the xylem and causing an 'air lock'.
2. The apparatus is set up as shown and then left under experimental conditions for 15 minutes, to reach equilibrium, before a bubble is introduced into the capillary tubing.
3. The rate of travel of the bubble (in cm/hour) is used to measure the rate of water uptake. If the bore of the tube is known, the rate of water uptake can be calculated in cm³/hour.

Using such an apparatus it is found that the environmental conditions that affect transpiration rate usually affect water uptake in the same way. Water uptake is increased by high temperature, low humidity and moving air, and decreased by low temperature, high humidity and still air. In the absence of light, water uptake decreases. These results suggest that transpiration is the major process controlling water uptake by the stem. This is, in fact, the case. Transpiration is the major factor governing water movement through the plant.

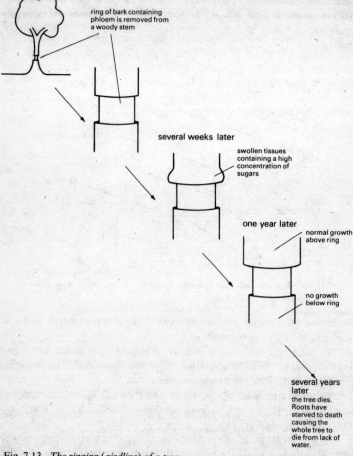

ring of bark containing phloem is removed from a woody stem

several weeks later

swollen tissues containing a high concentration of sugars

one year later

normal growth above ring

no growth below ring

several years later
the tree dies. Roots have starved to death causing the whole tree to die from lack of water.

Fig. 7.13. *The ringing (girdling) of a tree*

Uptake and transport of mineral salts

Dissolved mineral salts are absorbed into the plant as **ions**. The ions are taken up by diffusion in the osmotic flow of water through the root hairs. If needed in large quantities, particular ions are taken up by active transport, against the concentration gradient (see page 61). The plant may absorb one ion faster than another, depending on its needs (page 80).

On reaching the xylem, salts travel in the transpiration stream to the leaves and growing points. Here they are needed for protein synthesis.

'Ringing experiments' on stems, i.e. removing the bark and phloem (Fig. 7.13), show that transport of salts is unaffected, indicating that the salts do not travel through the phloem. The actual path taken can be shown by more sophisticated experiments using radioactive salts (for a similar experiment see page 160 and Fig. 7.14). The plant's roots are supplied with a radioactive isotope of a mineral, e.g. phosphorus. The path taken by phosphorus through the plant can be traced by sectioning, and using photographic techniques (see Fig. 7.14). The phosphorus is

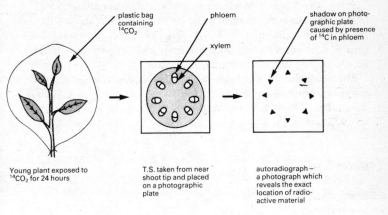

Fig. 7.14. *An experiment which demonstrates that organic compounds are transported to the stem phloem*

found in both xylem tissue and in the phloem. The experiment can be repeated this time inserting waxed paper as a barrier between the xylem and phloem. When this is done, the radioactive phosphorus is found only in the xylem vessels. These experiments indicate that phosphorus is transported in the xylem, but normally leaks across to the phloem. This sideways transport occurs through the pits in the xylem vessel walls.

7.10. *Transport in Phloem Sieve Tubes* (*Translocation*)

Translocation is the term usually used to describe the transport of organic substances from sites where they are made to other parts of the plant. The sugars and amino acids made in the leaf after photosynthesis (page 64), are transported in the phloem to growing parts and storage regions of the plant.

Definition: Translocation *is the transport of soluble organic substances within the plant.*

Movement of organic substances can occur up the stem to developing leaves, flowers and fruits, or down the stem into growing regions or storage organs in the root or stem.

The transport of organic substances in the phloem sieve tubes can be demonstrated in two sets of experiments, Figs 7.13 and 7.14. If a tree is 'ringed' by removing a ring of phloem-containing bark, then the tree will die within a season or so. Organic substances made in the leaves and transported down the stem cannot reach the roots and the roots eventually die.

NOTE. The flow of xylem from the roots up the stem is unaffected.

A second more recent experiment involves the use of carbon dioxide labelled with radioactive carbon (^{14}C). The ^{14}C is incorporated into organic substances during photosynthesis and the path then taken by the ^{14}C in the plant can be traced.

The mechanism of translocation through the phloem is not well understood. Transport depends on the activity of living cells in the phloem sieve tubes. If these cells are killed or their metabolic activity inhibited, translocation stops.

Definitions

Immunity
Transpiration
Translocation

Key Words

Circulatory system
Blood
Lymph
Arteries
Veins
Capillaries
Lymph vessels
Heart
Plasma
Platelets
  (thrombocytes)
Red blood cells
  (erythrocytes)
White blood cells
  (leucocytes)
Granulocytes
Lymphocytes

Monocytes
Thrombokinase
Prothrombin
Thrombin
Fibrinogen
Fibrin
Tissue fluid
Double circulatory
  system
Pulmonary
  circulation
Systemic circulation
Atria (auricles)
Ventricles
Systole
Diastole
Myogenic

Lymph nodes
Pathogens
Antibodies
Antigens
Active immunity
Passive immunity
Xylem
Phloem
Xylem vessels
Phloem sieve tubes
Vascular bundles
Lignin
Stomata
Root pressure
Capillarity
Potometer
Ions

Exam Questions

1. (*a*) *Make a large, fully labelled diagram to illustrate the internal structure of the mammalian heart and its associated blood vessels.* [10]
(*b*) *In the form of a table, briefly describe the differences between those vessels carrying blood away from the heart and those carrying blood towards the heart.* [6]
[O&C]

2. (*a*) *What is lymph?* [2]
(*b*) *Why does lymph appear milky after a fatty meal?* [2]
[OXF]

3. *Explain how blood clots and how a wound heals.* [4, 3]
[OXF]

4. *Describe how oxygen is taken up by the blood at the surface of the lung and passed from the blood to the respiring tissue outside the capillary.* [4, 5]
[OXF]

(See pages 127 and 128 before you answer this question.)

5. (a) *By means of labelled diagrams* only *show the distribution of xylem and phloem in a:*

(*i*)   *stem*

(*ii*) *root*

(*iii*) *leaf, of a* named *flowering plant.* [9]

(b) *What are the functions of xylem and of phloem?* [6]

(c) *Describe an experiment you could perform to show one of the functions of xylem.* [4]

(d) *How may external factors influence the rate at which xylem performs its functions?* [6]

[OXF]

6. (a) *Describe an experiment which would show that the rate of water loss from a leafy shoot is increased when the shoot is exposed to moving air.* [9]

(b) *Briefly describe the route taken and the processes involved as water passes into, through and out of a leaf.* [8]

[LON]

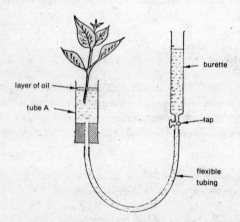

7. *The apparatus shown in the diagram can be used in the following experiment. (Methods of support have not been shown.)*

The level of the water in the tube A is marked and the time noted. After a suitable period of time (which is again noted) the tap is opened. Water is allowed to flow from the burette until the original level in tube A is regained. A reading from the burette reveals how much water is required to accomplish this.

(*a*) *What is the function of the layer of oil in tube A?*

(*b*) *The apparatus does not measure transpiration directly. What process does it measure?*

(*c*) (*i*) *If the apparatus is used to estimate the rate of transpiration, what assumption is being made?*

(*ii*) *Explain how you would use the results to calculate the rate of transpiration.*

(*d*) *Suggest* one *possible source of error in the apparatus as shown. State the effect this error may have on the results.* [7]

[AEB]

8. Keeping Conditions in the Body Constant (Homeostasis)

8.1. Homeostasis

Definition: Homeostasis *is the maintenance of a constant internal environment.*

In practice, when dealing with a mammal, homeostasis means keeping the **blood** constant in composition and temperature. Remember, it is blood which gives rise to the tissue fluid which bathes the cells in the body (see page 144).

The advantage of homeostasis is that it makes the functioning of the body independent of changes in the external environment. The cells can be kept at their optimum conditions whereby they will function most efficiently. Any variations in pH or temperature can have a seriously inhibiting effect on enzyme action and will lower metabolism (see page 49).

Homeostasis is an active process and requires energy.

In a mammal, the following features of tissue fluid must be kept constant:

1. the balance of chemical constituents such as salts, dissolved foods and waste products
2. the osmotic potential (O.P.), as determined by the relative amounts of water and dissolved substances (page 58)
3. the level of carbon dioxide
4. temperature.

In general, we recognize four major sets of organs (homeostatic organs), which regulate these features.

1. the **liver**, which, for example, regulates blood glucose level (page 96)
2. the **lungs**, which regulate the carbon dioxide level in the blood (page 122)
3. the **kidneys**, which regulate the level of salt and water in the blood, and hence regulate the blood's O.P.; they also control the level of nitrogenous wastes and poisons in the blood
4. the **skin**, which regulates body temperature by increasing or decreasing heat loss through the body surface.

Two important aspects of homeostasis are **excretion** and **osmoregulation**.

Definition: Excretion *is the removal from the body of the waste products of metabolism.*

Excretion should *not* be confused with egestion (page 82) or secretion (page 82).

Definition: Osmoregulation *is the process by which the osmotic potential (O.P.) of body fluids is kept constant.*

8.2. *Excretion*

The removal of waste products of **metabolism** is a vital part of homeostasis. It prevents the accumulation of substances which are poisonous or would otherwise interfere with the normal functioning of body cells. These excreted waste substances are called **excretory products**.

The term excretion has also come to include the removal from the body of excess or poisonous substances which have been absorbed. In the case of water, for example, it is not possible to tell whether excreted water has come from metabolism (e.g. respiration) or whether it has been absorbed.

The major excretory products and excretory structures in mammals and flowering plants are shown in Table 8.1. Refer to this table and the page references when answering the following question.

Q. (a) What is excretion? [2]
(b) Name the products excreted by
(i) flowering plants and
(ii) mammals [6]
(c) Where and how are these excretory products formed? [8]
[LON]

Table 8.1. *The excretory products and excretory structures of mammals and flowering plants*

	Mammals	Flowering plants
	Mammals excrete gases or dissolved solids	Plants excrete gases or undissolved solids
Excretory products:	Carbon dioxide (from respiration)	Carbon dioxide (from respiration)
	Water (from respiration or excess absorbed)	Water (from respiration)

	Mammals	*Flowering plants*
	Nitrogenous wastes, e.g. urea (from protein metabolism)*	Oxygen (from photosynthesis)
	Bile pigments (from breakdown of red blood cells)	Plants excrete few nitrogenous wastes. They make their own proteins and do not have an excess to break down and remove
	Mineral salts (excess of those absorbed)	
Excretory structures:	Lungs – excrete carbon dioxide and water vapour (see page 127)	Stomata – excrete oxygen, carbon dioxide and water vapour (see page 129)
	Kidneys – excrete water, salts, nitrogenous wastes and certain poisons (see page 168)	Whole leaves – some deciduous trees, e.g. oak, transfer wastes to the leaves which are then shed in autumn
	Skin – excretes salts, water and urea from sweat glands (see page 175)	Heartwood – some solid wastes are deposited in the wood at the centre of tree stems
	Liver – excretes bile pigments into the duodenum	Cell vacuoles – some solid wastes are deposited in the vacuoles of living cells, e.g. silicates in the leaves of grasses

Students commonly confuse excretion with egestion. Mammalian faeces contain mainly *egested* material, that is, substances which have not been digested and have not been absorbed. These substances have never passed across a cell membrane into the body and so cannot be excreted. The faeces contain only a small amount of excreted material such as bile salts.

Insects (page 335) are different in this respect. Their faeces contain a large proportion of excreted material. Their Malpighian tubules (roughly equivalent of kidneys) excrete the nitrogenous waste, uric acid, into the gut. It is removed from the body as part of the faeces.

Unicellular animals and plants such as *Amoeba* (page 311) and *Chlamydomonas* (page 326), and the smaller multicellular animals such as *Hydra*

* Animals other than mammals secrete nitrogenous wastes other than urea (see page 168).

(page 328), do not need special excretory organs. Their surface area is large in relation to their volume and waste products can simply diffuse out of the body as quickly as they are formed.

8.3. *Osmoregulation*

Both plants and animals have the problem of maintaining their body fluids at a relatively constant osmotic potential (page 58). However, animals have two problems which plants do not have:

1. **Animal cells do not have cell walls:** if water enters by osmosis the cells must either be surrounded by liquid of the same O.P., as in multicellular organisms, or the cells must have special mechanisms to remove excess water, as in unicellular organisms.

In **freshwater**, body fluids have a higher O.P. than the surrounding water, and water will flow into the body across any semi-permeable surface. In *Amoeba* (page 311), excess water flowing in by osmosis is pumped out by a contractile vacuole (Fig. 8.1). This requires energy from

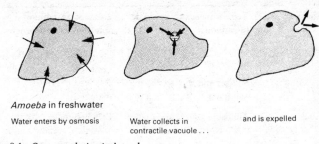

Amoeba in freshwater

Water enters by osmosis   Water collects in   and is expelled
                                 contractile vacuole . . .

Fig. 8.1. *Osmoregulation in* Amoeba

respiration. In bony fish (page 337), most of the body surface is impermeable because of its layer of scales and mucus. Water enters by osmosis across the gill epithelium and is also absorbed through the gut wall. Excess water is excreted in urine produced by the kidneys.

In **seawater**, the body fluids of invertebrates (page 304) have about the same O.P. as the surrounding seawater. Hence an *Amoeba* capable of living in both sea and freshwater will have a contractile vacuole in freshwater, but will lose it in seawater. The situation in vertebrates is more complex. The body fluids of bony fish have an O.P. slightly *below* that of seawater. They tend to *lose* water by osmosis. They get round this problem by drinking seawater and pumping out excess salts across their gills. They produce small volumes of concentrated urine.

Terrestrial (land) animals have the added problem of water loss by evaporation.

2. **Animals produce nitrogenous wastes** and water is used in their removal from the body. This can be a source of serious water loss.

Remember, nitrogenous (nitrogen-containing) wastes are derived from the breakdown of unusable proteins and amino acids.

$$\text{protein} \rightarrow \text{amino acids} \rightarrow \text{amine } (NH_2) \begin{array}{l} \nearrow \text{ uric acid} \\ \rightarrow \text{ ammonia} \\ \searrow \text{ urea} \end{array}$$
$$\text{group}$$
(see Table 8.2)

With the exception of mammals, animals living in dry environments where there is a serious problem of water loss, tend to excrete their nitrogenous waste in a semi-solid form. Mammals and aquatic animals excrete nitrogenous wastes dissolved in water.

Table 8.2. *Types of nitrogenous waste*

Nitrogenous waste	Details	Animal
Ammonia	Very soluble and very toxic (poisonous)	Excreted as a dilute solution by most aquatic animals other than mammals, e.g. *Amoeba* and bony fish
Urea	Soluble and less toxic than ammonia	Excreted in urine by mammals
Uric acid	Almost insoluble and non toxic. Normally excreted as a white paste	Excreted by terrestrial animals, e.g. insects and terrestrial vertebrates other than mammals

8.4. *The Mammalian Kidney*

The kidneys are the major organs in a mammal concerned with **excretion** and **osmoregulation**. The kidneys remove the following substances from blood:

1. soluble nitrogenous wastes, mainly urea
2. excess water
3. excess salts
4. excess hydrogen ions (to keep blood pH constant)
5. foreign substances such as drugs, poisons, etc.
6. hormones.

All these functions are excretory, and in addition, 2 and 3 are osmo-regulatory.

The above substances are excreted by the kidneys in the **urine**.

Kidney structure

The mammalian kidneys are a pair of bean-shaped, reddish-brown organs enclosed in a transparent membrane. They are attached to the back of the abdominal cavity (Fig. 8.2).

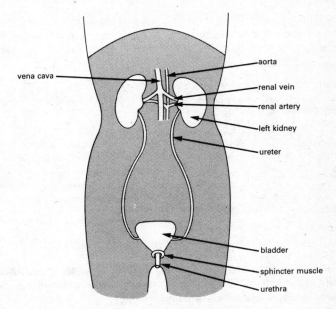

Fig. 8.2. *The mammalian kidneys and associated structures*

The kidneys are connected to other parts of the body by three vessels:
1. the **renal artery**, which delivers blood to the kidney from the aorta
2. the **renal vein**, which carries blood away from the kidney to the vena cava
3. the **ureter**, which carries urine away to the bladder.

In vertical section (Fig. 8.3) the kidney consists of a dark outer region, the **cortex**, and a lighter inner region, the **medulla**. The inner border of the medulla forms several conical **pyramids** which open out into the **pelvis** of the kidney. The pelvis extends into a tube, the ureter.

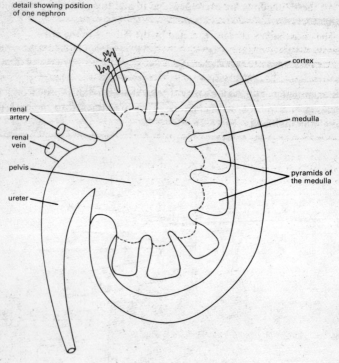

detail showing position
of one nephron

cortex

renal
artery

medulla

renal
vein

pelvis

pyramids of
the medulla

ureter

Fig. 8.3. *V.S. of the mammalian kidney*

Invisible to the naked eye are the tiny **nephrons** which are the functional units of the kidney. The position of one of these is shown in Fig. 8.3. There are about one million in a human kidney.

Each nephron consists of a bundle of capillaries, the **glomerulus**, surrounded by a **Bowman's capsule**, which empties into a **kidney tubule** (see Fig. 8.4). The tubule is divided into three regions, the **first** (proximal) **convoluted tubule**, the **loop of Henlé** and the **second** (distal) **convoluted tubule**. The tubule leads into a collecting duct which empties into the pelvis of the kidney.

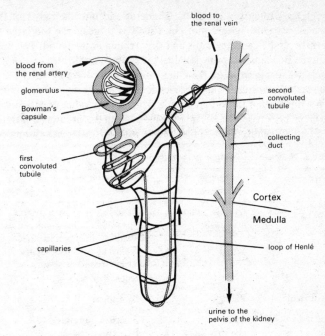

Fig. 8.4. *A nephron from the mammalian kidney*

Kidney function

The mammalian kidneys work by filtering the blood, taking back (re-absorbing) from the filtrate those substances which are valuable, and allowing those substances which are harmful or in excess, to leave the body in the urine.

This mechanism involves two processes:

1. **pressure** or **ultra-filtration**
2. **selective reabsorption**.

1. **Pressure filtration** (in the Bowman's capsule): blood enters the kidney along the renal artery which branches into many arterioles. One arteriole leads into the glomerulus of each nephron. This arteriole is larger than the one leaving the glomerulus, so that within the glomerulus the blood is at high pressure. This forces plasma and smaller dissolved substances

through the capillary walls of the glomerus and into the cavity of the Bowman's capsule. Larger molecules such as proteins are too large to pass through the capillary walls and they remain in the blood. The fluid within the Bowman's capsule is called the **glomerular filtrate**.

2. **Selective reabsorption** (in the kidney tubule and collecting duct): from the Bowman's capsule the arteriole leading out breaks up into capillaries which wrap around the kidney tubule. This enables valuable substances to be taken back (reabsorbed) into the blood from the glomerular filtrate. The different regions of the kidney tubule reabsorb different substances.

Table 8.3. *Selective reabsorption in the kidney*

Region	Substances reabsorbed	Mechanism notes
First convoluted tubule	Glucose, amino acids, salts	Taken up by active transport (page 61) against a concentration gradient. Energy is required
	Water	Taken up by osmosis (page 57). Over 80 per cent of water is reabsorbed here
Loop of Henlé	Water	By osmosis
Second convoluted tubule	Salts	By active transport
	Hydrogen ions	By active transport
	Water	By osmosis*
Collecting duct	Water	By osmosis*

The excretory substances and excess water left in the collecting ducts empty into the pelvis as urine.

Urine

From the pelvis, the urine drips into the ureter which empties into the muscular sac, the bladder, where the urine is stored. Under voluntary control, muscular contractions of the bladder squirt urine into the **urethra** and out of the body. (Do not confuse the ure**thra** with the ure**ter** above.)

A human produces about 1·5 litres of urine per day on average. The urine is about 96 per cent water, 2 per cent urea and 2 per cent salts; it also contains hormones (variously modified) and possibly some poisons or drugs in small amounts. The exact composition of urine varies according to diet, activity and health. For example, people suffering from the

* Uptake of water in the second convoluted tubule and collecting duct is controlled by the hormone ADH according to the body's need to conserve water.

condition *diabetes mellitus* lose excess glucose in their urine (glucose is not a normal constituent of urine).

The volume and composition of urine varies according to the body's need to conserve water. In hot weather, the body loses water by sweating (page 176), so to conserve water the kidneys reabsorb as much as possible. As a result, a small volume of concentrated urine is produced.

After drinking large volumes of liquid the O.P. of body fluids may become lowered. In forming urine, the kidneys will reabsorb as little water as possible, and a large volume of dilute urine will be produced.

The amount of water lost in the urine is controlled by the hormone ADH (see page 203).

8.5. *The Mammalian Skin*

Skin covers the body surface. It is described as an organ because it is made up of different tissues.

Skin structure

The skin has two main layers:
1. an outer layer, the **epidermis**
2. an inner layer, the **dermis**.

1. **Epidermis:** the epidermis is the protective surface layer many cells thick. Dead cells on top are constantly worn away and replaced by newly formed cells in the **germinative layer**. As cells move towards the surface **keratin** is deposited in them. The cells became flattened, lose their nuclei and die. Keratin waterproofs the surface and reduces water loss from tissues below.

Downfolds of the epidermis form the sweat glands, sebaceous glands and hair follicles, in the dermis.

2. **Dermis:** the dermis contains connective tissue, hair follicles, blood vessels, glands, muscles, nerves and sense organs.

The **hair follicle** secretes the non-living hair. Attached to the hair is an **erector pili muscle** which, when it contracts, pulls the hair upright.

Sebaceous glands open into the hair follicle and secrete an oily fluid, **sebum**, which helps keep the hair and skin surface supple, waterproof and bacteria-free.

Sweat glands open on to the skin surface via sweat ducts. They secrete sweat, a fluid drawn from the blood.

Blood vessels in the dermis bring food and oxygen to the skin tissues.

Some give rise to the fluid, sweat, and some form capillary loops. Both play an important part in temperature regulation (section 8.6).

Five types of **sensory receptor** in the skin detect heat, cold, touch, pain and pressure. **Motor nerves** (page 182) stimulate muscles and glands, while **sensory nerves** (page 182) run from the skin sensory receptors to the brain.

A fat layer (adipose tissue) below the dermis protects underlying organs from mechanical damage and acts as a food store and insulator.

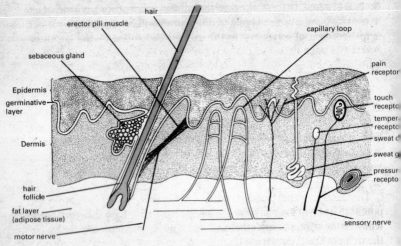

Fig. 8.5. *V.S. of mammalian skin*

Skin function

The skin's functions relate to it being the boundary layer of the body. Its functions are:

1. **Protection:**
(a) The outer dead layer of the epidermis is impermeable and prevents loss of water.
(b) The skin protects against mechanical damage and friction.
(c) A dark pigment **melanin** in the germinative layer prevents u.v. light (from sunlight) penetrating the body and damaging vital organs.
(d) The skin prevents entry of micro-organisms (page 147), the sebum kills bacteria.
(e) The fat layer and hairs reduce heat loss.

2. **Temperature regulation:** sweating, together with widening (dilation) and narrowing (constriction) of blood vessels, controls heat loss through the skin (see page 176).

3. **Vitamin D production:** in the presence of u.v. light (in sunlight) certain fats are converted to vitamin D. Vitamin D is otherwise provided by the diet.

4. **Sensitivity:** the skin is sensitive to touch, heat, cold, pain and pressure. These are protective functions, enabling the body to respond to adverse conditions; they also provide considerable information about the body's surroundings.

5. **Excretion:** water, urea and dissolved salts are excreted in the sweat. The loss of these substances is incidental. The main function of sweat is to increase heat loss.

NOTE. Structures such as teeth (page 98), mammary glands (page 256), nails, hairs, hooves and claws are modified skin tissues.

8.6. *Control of Body Temperature*

Mammals and birds are **homoiotherms**, meaning they have a constant body temperature. Popularly, they are described as 'warm-blooded'. Other vertebrates and all invertebrates are **poikilotherms**, meaning they have a variable body temperature. They are commonly described as 'cold-blooded'.

The terms 'warm-blooded' and 'cold-blooded' are unreliable. A poikilothermic organism living in a hot desert may have a higher body temperature than a homoiothermic organism. It is wiser therefore to use the technical and not the popular terms.

The body temperature of a poikilotherm is largely governed by the environmental temperature and the degree of activity of the organism. In cold weather the body temperature of a poikilotherm will drop and the animal may become sluggish and inactive. A homoiotherm regulates its body temperature so that it is largely independent of external temperature. This has two major advantages:

1. A homoiotherm can maintain the optimum termperature for chemical reactions in the body (metabolism).

2. A homoiotherm can be active all the year round and at all times of day, whatever the external temperature.

In man, the body temperature is kept at 37 °C, which is the optimum temperature for enzyme action. Death occurs if the body temperature rises much above 42 °C or drops much below 30 °C.

Homoiothermic organisms such as man usually have a body temperature higher than their surroundings. To maintain their temperature they must generate their own body heat, largely from respiration. To keep this temperature constant they need to balance the heat they gain against the heat they lose so that:

heat gained by body = heat lost from body

The body gains heat from:

1. metabolism – **respiration** in the liver and muscles releases a large amount of heat energy
2. heat **radiation** from the sun or nearby hot objects
3. heat **conduction** from the air or any hot objects in contact with the body (including hot food or drink which is ingested).

The body loses heat by:

1. **evaporation** of sweat from the skin or water from the lungs (NOTE. **Latent heat** is needed to convert a liquid into a gas)
2. heat **radiation** to the surroundings
3. heat **conduction** to the surrounding air or any cool objects in contact with the body
4. heat loss by **convection** (warm air next to the body rises and is replaced by colder air)
5. some heat is lost in urine and faeces.

A small structure, the **hypothalamus**, at the base of the fore-brain (page 188) acts as a thermostat, monitoring blood temperature. If the temperature rises above normal, nerve impulses are relayed to the sweat glands, blood vessels and erector muscles in the skin to increase heat loss. If blood temperature drops below normal, the hypothalamus controls an appropriate set of responses to reduce heat loss and increase heat production. The responses to overheating and overcooling are shown below. The control mechanism is summarized on page 204.

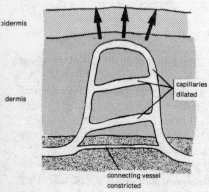
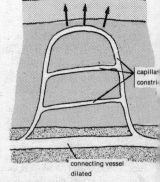

Fig. 8.6.

(a) Vasodilation   (b) Vasoconstriction

1. **Control of overheating** (blood temperature in man rises above 37 °C): body temperature may rise during vigorous exercise, when the body is fighting infection, or if the air temperature is high. The following responses, controlled by the hypothalamus, lower body temperature:

(a) **Increased sweating:** sweat glands form sweat from the water, salts and urea they receive from blood capillaries. As the water in sweat evaporates from the skin's surface, so the skin is cooled.

(b) **Vasodilation** (widening of capillaries in the dermis, see Fig. 8.6): the blood flow to the skin is increased and heat loss by radiation, convection and conduction is increased.

(c) **Reduced muscular activity** (NOTE. Respiration supplies muscles with the energy required for contraction but also produces heat as a by-product).

(d) **Hair lies flat against the skin:** not very important in man (see below).

(e) **Voluntary actions** may be employed such as removing clothes, reducing artificial heating or switching on a fan.

(f) **Reduced rate of metabolism:** in the long term the appetite may be reduced and less food broken down in respiration or stored.

Failure of the above mechanisms to lower body temperature may result in heat stroke, **hyperthermia**, and then heat death.

2. **Control of overcooling** (blood temperature drops below 37 °C):

(a) **Sweating stops**.

(b) **Vasoconstriction** (narrowing of blood vessels in the dermis, (see Fig. 8.6). The blood flow to the skin is decreased and heat loss by radiation, convection and conduction is reduced.

(c) **Increased voluntary muscular activity** increases heat production.

(d) **Increased involuntary muscular activity** (shivering) increases heat production.

(e) **Hair stands on end** ('goose flesh'): this is of little importance in man, but in other animals with a thick covering of hair (fur) the air trapped between erect hairs significantly reduces heat loss by radiation, convection and conduction.

(f) **Voluntary actions** such as putting on additional clothes or increasing artificial heating may be employed.

(g) **Increased rate of metabolism:** in the long term appetite may be increased and more food broken down by respiration in the liver or stored as fat for insulation.

Failure of the above mechanisms to raise body temperature may result in reduced 'deep body' or 'core' temperature, **hypothermia**, and then cold death.

Definitions

Homeostasis
Excretion
Osmoregulation

Key Words

Liver
Lungs
Kidneys
Skin
Stomata
Excretory products
Nitrogenous wastes
Bile pigments
Ammonia
Urea
Uric acid
Contractile vacuole
Malpighian tubules
Cortex
Medulla
Ureter
Renal artery
Renal vein
Nephron
Pyramids
Pelvis
Glomerulus

Bowman's capsule
Glomerular filtrate
Kidney tubule
First convoluted
  tubule
Loop of Henlé
Second convoluted
  tubule
Collecting duct
Urine
Pressure (ultra)
  filtration
Selective reabsorption
Bladder
Urethra
Diabetes mellitus
ADH (anti-diuretic
  hormone)
Epidermis
Dermis
Keratin
Germinative layer

Hair follicles
Sweat glands
Sebaceous glands
Sebum
Erector pili muscles
Sense organs
Capillary loops
Homoiothermic
Poikilothermic
Radiation
Conduction
Convection
Evaporation
Hypothalamus
Sweating
Latent heat
Vasodilation
Vasoconstriction
Hyperthermia
Hypothermia

Exam Questions

1. (a) What is excretion?   [2]
   (b) Name the products excreted by
   (i) flowering plants, and
   (ii) mammals.   [6]
   (c) Where and how are these excretory products formed?   [8]
                                                              [LON]

2. Which of the following best describes the process in mammals by which liquid passes from the blood capillaries into the kidney tubules?

A. *Osmosis*

B. *Diffusion*

C. *Excretion*

D. *Filtration*

E. *Reabsorption* [1]

[LON]

3. (a) What is the importance of urea and where is it formed in the mammalian body? [4]

(b) Draw and label a large generalized plan to show the urinary (excretory) system of a mammal and its associated blood vessels. [6]

(c) Give an account of how urine is formed within the kidney. [6]

[O&C]

4. The diagram below shows a mammalian kidney tubule from Bowman's capsule to one of the collecting ducts.

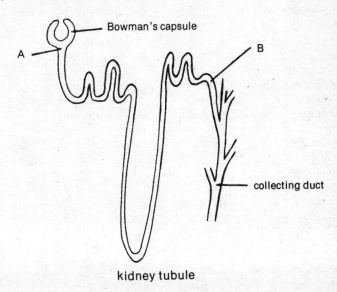

kidney tubule

(a) Describe three differences between the contents of the tubule at **A** and at **B**.

(b) In which part of the kidney would you expect to find

(i) Bowman's capsules?

(ii) collecting ducts? [5]

[AEB]

5. (a) *Make a labelled diagram of a section through the mammalian skin.*
[6]

(b) *For a* named *species of mammal briefly describe how:*
(i) *water is gained and lost* [7]
(ii) *heat is gained and lost.* [5]
(c) *How are these changes regulated by the mammal?* [7]

[OXF]

(For part (c) see pages 203 and 204.)

9. Control and Coordination

Definition: Sensitivity (irritability) *is an organism's ability to respond to a stimulus.*

External stimuli include changes in temperature, light and sound. The response often takes the form of some kind of movement. In general, responses are much more rapid in animals than in plants.

Coordination is the linking together of bodily functions in space and time. It also involves the ability to respond to changes in the external environment. The processes we have examined in earlier chapters – nutrition, respiration, excretion, etc. do not take place independently of one another, but are interrelated. For example, aerobic respiration in a contracting muscle would soon cease if the lungs failed to supply oxygen via the circulatory system. During exercise, when the muscles need to gain more oxygen and glucose, and remove excess carbon dioxide, breathing rate is automatically increased and the heart beats faster. This sends a greater volume of oxygenated blood to the muscles. Without coordination, bodily activities would soon malfunction; for example, a runner would collapse after a few metres through lack of an increased oxygen supply to his/her muscles.

The activities of organs and organ systems are not only closely linked to each other and the overall pattern of activity within the body, but to changes outside the body. In response to environmental changes the mammal makes adjustments that tend to keep its internal conditions constant, i.e. homeostasis (page 164). Thus, patterns of activity in the body are responsive to external as well as internal changes.

In mammals and other multicellular animals, coordination is brought about by the action of the **nervous system** (involving electrical **nerve impulses**) and the **endocrine system** (involving chemical messengers called **hormones**). In plants, coordination is under the control of hormones.

9.1. *The Mammalian Nervous System*

The mammalian nervous system is divided into two parts, the **central nervous system** (C.N.S.) – the brain and spinal cord – and the **peripheral nervous system** – the nerves which run into and out of the C.N.S.

Information is transmitted through the nervous system in the form of nerve impulses – essentially weak electric currents. Stimuli from the internal or external environment stimulate **sensory receptors** which convey nerve impulses along **sensory nerve cells** into the C.N.S. The C.N.S. 'processes' this information and may then deliver nerve impulses along **motor nerve cells** to **effectors**, which give an appropriate response. Examples of sensory receptors are sense organs such as the eye (page 189) and the ear (page 195) and various sensory cells such as pressure receptors in the skin (page 174) and stretch receptors in the muscles (page 224). Muscles and glands are effectors – structures which give a response.

The functional units of the nervous system are nerve cells (neurones). There are three types:

1. **Sensory neurones** convey nerve impulses from sensory receptors to the C.N.S.

2. **Relay or intermediate neurones** transmit nerve impulses within the C.N.S.

3. **Motor neurones** convey nerve impulses from the C.N.S. to effectors.

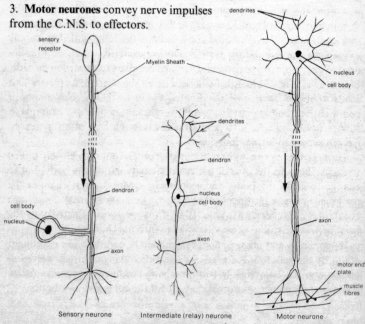

Fig. 9.1. *Three types of nerve cell (neurone)*

These three types of cell are illustrated in Fig. 9.1. You should learn these diagrams well – questions on sensory or motor neurones often come up. The function of these cells is to transmit a nerve impulse from one end of the cell to the other. The neurone is elongated so that the nerve impulse can be conveyed a long distance. This elongated part of the nerve cell is called the **nerve fibre**. In comparing sensory and motor neurones notice the position of the cell body and the absence of dendrites in the sensory neurone. Parts of the cell beginning with **den-** convey a nerve impulse to or past the cell body. The **axon** is that part of the nerve fibre which conveys a nerve impulse away from the cell body.

A nerve impulse is passed on from one cell to the next across a small gap called a **synapse**. The synapse allows transmission in only one direction so, as a rule, nerve impulses travel along particular nervous pathways in one direction only. When a nerve impulse reaches a synapse it causes the release of small amounts of a chemical **transmitter** substance, which triggers a nerve impulse in the adjacent neurone. Synapses are also found between the motor end plate of a motor neurone and an effector. A chemical transmitter released across the synapse stimulates activity in the effector, e.g. contraction of a muscle, or secretion by a gland. Some transmitter substances also act as hormones, e.g. adrenalin (page 200).

Activities within the mammalian nervous system are incredibly complex and only the very simplest stimulus–response reactions can easily be described. Reflex actions are such examples.

9.2. *Reflex Action*

Simple reflexes

Definition: A reflex action *is a rapid automatic* (*involuntary*) *response to a stimulus.*

Examples of reflexes in man are coughing, sneezing, blinking, sweating, the knee jerk reflex and removing a hand from a hot object. Reflex actions are involved in **homeostasis**, i.e. keeping conditions in the body constant, and in preventing damage to the body. For example, changes in breathing rate during exercise occur automatically without conscious control. This reflex action is a homeostatic function: it serves to keep the blood carbon dioxide level approximately constant. The rapid removal of a hand from a hot object occurs without conscious control. If you waited to 'think' about what response to give, your hand could well be burnt. Instead, the brain is bypassed and preventative action taken before the brain 'decides'.

In reflex action, the nervous pathway between receptor and effector is known as a reflex arc. Simple reflex arcs normally have five components: a sensory receptor, a sensory neurone, an intermediate neurone, a motor neurone, and an effector. The reflex arc for the hand withdrawal reflex is shown in Fig. 9.2.

Study this diagram carefully: in an exam you might have to draw this,

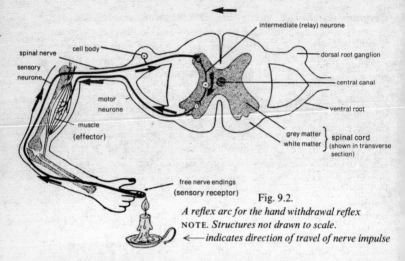

Fig. 9.2.
A reflex arc for the hand withdrawal reflex
NOTE. *Structures not drawn to scale.*
← *indicates direction of travel of nerve impulse*

or a similar reflex arc, or you may have to label or comment on one provided. Note the positions of neurones and their cell bodies. The sensory neurone runs through the dorsal root of the spinal cord whereas the motor neurone runs through the ventral root. The grey matter of the spinal cord contains a high concentration of cell bodies (it is the nuclei which make grey matter 'grey'), while the white matter contains nerve fibres. You should be aware that the diagram is a simplification. Hundreds of reflex areas are involved in the reflex withdrawal of a hand from a hot object.

Reflexes which involve the spinal cord only and not the brain, are called **spinal reflexes**. Although the brain is not directly involved in these reflex arcs, it does not mean the brain is not informed of events. Although your hand may be removed automatically from a hot object, a short time later you will feel the pain and may let out a cry. The intermediate neurone in the reflex arc has synapses with other intermediate neurones which convey nerve impulses up the spinal cord to the brain. Also, nerve impulses can be sent down from the brain via intermediate neurones to modify reflex action. For example, the simple reflex response to picking up a very hot object might be to drop it. However, if the object is a pan of boiling water, the person may place the pan down safely rather than spill the contents all over him- or herself. Impulses from the brain have overridden the simple reflex.

Connections between reflex arcs and the brain enable reflex actions to be modified by experience. Conditioned reflexes are examples of such modifications.

Conditioned reflexes

In most simple reflexes, the stimulus and response are related. For example, the chemical stimulus of food in the mouth triggers the salivary reflex (secretion of saliva). After a period of learning or training, however, it is possible for a different and often irrelevant stimulus to produce the same response. In such a case a 'conditioned reflex' has been set up and the animal is said to be conditioned to the new stimulus. The classic experiments on conditioned reflexes were carried out by the Russian biologist Pavlov at the beginning of this century.

In dogs, the salivary reflex is triggered by the smell or taste of food. For several days, Pavlov rang a bell at the time food was given to experimental dogs. Later, the sound of the bell alone was a sufficient stimulus to cause a dog's mouth to water, without sight or smell of the food. The original chemical stimulus of the food has been replaced by an unrelated stimulus detected by the ears.

The training of animals is done largely by conditioning them to respond to new stimuli. To what extent conditioned reflexes play a role in human behaviour is difficult to say. However, the learning of habits and basic skills probably occurs in a similar fashion. For example, walking, running, cycling, using a knife and fork, etc., although all originally learned consciously, later function below the level of consciousness. However, these patterns of behaviour are clearly much more complex than simple conditioned reflexes.

The advantage of simple and conditioned reflex responses is that they leave the conscious mind free to think about other less routine matters which do require concentration and reasoning.

9.3. *The Central Nervous System*

The spinal cord

The spinal cord is a hollow tube comprising white matter (nerve fibres) on the outside, grey matter (cell bodies) in an H-shaped zone on the inside, and a central canal containing nutritive fluid (see Fig. 9.2 above); memorize this arrangement.

The spinal cord runs through, and is protected by, the vertebral column (see page 217). At regular intervals paired, spinal nerves emerge from the spinal cord. These nerves typically contain both sensory and motor nerve fibres (see Fig. 9.2 above) enclosed in a connective tissue sheath.

NOTE. A **nerve** contains many hundreds of **nerve fibres** – do not confuse the two.

The spinal cord has three main functions:

1. It contains the circuitry for reflex arcs involving the muscles or glands of the body, as distinct from those of the head.
2. It provides connections between reflex arcs, both along and across the cord, so that a limited degree of coordination is possible.
3. It acts as a relay station for impulses passing between the brain and the lower parts of the body.

Brain

The brain is the enlarged front end of the spinal cord in vertebrates. It acts as a centre for reflex arcs in the head involving the main sense organs, the eyes, ears, mouth and nose. The increased importance of these sense organs has led, during evolution, to an increase in the size and complexity of the brain.

Like the spinal cord, the brain consists of nerve cells with a great concentration of cell bodies. In simpler vertebrates such as fish and amphibians, three regions of the brain are distinguishable: the **fore-**, **mid-** and **hind-brain**. The fore-brain receives nerve impulses from the nasal organs, the mid-brain receives impulses from the eyes, while the hind-brain receives impulses from the ears and sensory receptors in the skin. In mammals, these three regions are not clearly distinguishable due to the great expansion of the fore-brain to form the **cerebral hemispheres** or **cerebrum**. In man, this is the largest part of the brain and spreads out and covers most of the brain.

The major parts of the human brain are as follows (see Fig. 9.3).

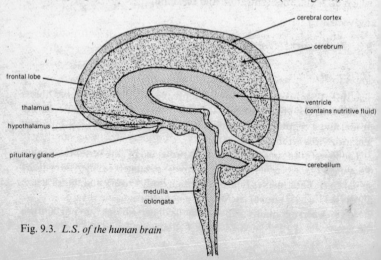

Fig. 9.3. *L.S. of the human brain*

1. Derived from the **hind-brain**:

The **medulla oblongata** is the central part of the autonomic nervous system – this system coordinates internal and mainly unconscious (involuntary) activities such as regulation of body temperature, blood pressure, and rates of heartbeat and breathing.

The **cerebellum** is a reflex centre which receives impulses from the organs of balance in the inner ear (page 195) and from stretch receptors (page 224) in joints and muscles. This information is used to achieve balance and muscular coordination in activities like walking, running and riding a bicycle.

2. Derived from the **mid-brain**:

The **thalamus** is a relay centre, directing sensory nerve impulses to appropriate parts of the cerebral hemispheres (see below).

3. Derived from the **fore-brain**:

The **cerebral hemispheres** have a thin outer layer of grey matter (cell bodies) called the **cerebral cortex**, which overlaps white matter (nerve fibres). The cerebral cortex is folded so that it has a far greater mass of grey matter than it would if it had a smooth surface.

All parts of the cortex have nerve cells of similar appearance, yet different parts of the cortex have different functions. Using stimulation experiments (applying mild electric shocks to different parts of the brain) and by observing the effects of brain damage in accident victims, it is possible to plot a map of the brain's surface to show the functions of different areas. One region of each cerebral hemisphere is called the **motor area**, because shocks applied here result in contraction of muscles in various parts of the body. These regions have connections which pass down the spinal cord to motor neurones. Motor areas direct conscious movement of the body.

If other areas of the cortex are electrically stimulated, patients experience noises, sight and sensations of touch. These **sensory areas** receive impulses from sense organs all over the body. There are separate sensory areas for vision, hearing, touch, taste and smell.

Other areas of the cortex are 'silent' when stimulated. These silent or **association areas** are where 'association' takes place between information from all sense organs, together with remembered information from past experience, to produce conscious awareness and understanding of the outside world. Appropriate responses can be made based not only on present sensory information, but also on past experience. Without such association areas, learning and conditioned reflexes would not be possible.

The frontal lobes of the cerebral hemispheres appear to be responsible for a person's individuality, character, imagination and intelligence.

The **hypothalamus** is a small downfold of the fore-brain and is a reflex centre concerned with many homeostatic functions. It has a direct nervous connection with the pituitary gland, and in many cases exerts a controlling influence by stimulating or inhibiting secretion of pituitary hormones (see pages 199–201).

The functions of the brain, in summary:

Most of the sensory information which enters the C.N.S. passes through the brain, which has control over most of the responses made by the body (voluntary action). Using association areas, impulses from all sense organs are integrated and then, via motor areas, coordinated responses are made. The brain is also an organ of memory, learning and reasoning. Stored information can be used to modify behaviour according to past experience.

9.4. *Sensory Receptors: Sensory Cells and Sense Organs*

Sensory receptors are specialized cells which convert stimuli into nerve impulses. These cells are the point of entry for information entering the nervous system. Sensory receptors may consist of isolated single cells, e.g. pressure receptors in the skin (page 174), or they may be collected together to form a highly efficient sense organ, e.g. the eye (page 189).

In mammals, it is only in the brain that a stimulus is interpreted (perceived) and identified. For example, if the nerve running from the eye is damaged, the person may be blind although the eye itself may function perfectly.

1. **External receptors** detect stimuli arising from outside the body. The main ones are:

(a) **Skin receptors**, which detect heat, cold, touch, pressure and pain (see Fig. 8.5, page 174). Certain regions of skin are more sensitive to particular stimuli than others. For example, the tip of the tongue, the lips and fingertips are very sensitive to touch. In these areas the appropriate sensory cells each have their own nerve fibre running to the C.N.S. In other less sensitive areas, several sensory cells share the same nerve fibre.

(b) In the nasal cavity, **olfactory organs** are sensitive to smell. They detect chemical substances which dissolve in the film of moisture covering the sensory cells.

(c) On the tongue, **taste receptors** are grouped together in **taste buds** which, like olfactory organs, detect chemical substances which dissolve in the film of moisture covering them. There are four types of taste bud sensitive to bitterness, saltiness, sourness and sweetness.

(d) In the retina of the eye, **light receptors**, rods and cones, are sensitive to different wavelengths and intensities of light (see page 194).

(e) In the inner ear (see page 196), the **organ of Corti** in the cochlea is sensitive to vibrations (sound). **Sensory hair cells** in the sacculus, utriculus and semi-circular canals are sensitive to the orientation and movement of the body.

2. **Internal receptors** respond to stimuli within the body and are often associated with homeostatic mechanisms (see pages 202–4);

(a) the **hypothalamus** detects changes in blood temperature (see page 204).

(b) The **hypothalamus** detects changes in blood osmotic potential (see page 203).

(c) The **respiratory centre** of the medulla oblongata is sensitive to carbon dioxide level (see page 202).

(d) **Proprioreceptors** respond to stimuli from muscles and joints. One type, stretch receptors (page 224), are embedded in every skeletal muscle in the body, and similar receptors are found in muscle tendons, and in the ligaments of joints (page 215). These receptors provide the brain with information about the degree of tension in each part of the muscular system, and the angle to which every joint is bent. Without proprioreceptors, muscular coordination and locomotion would be impossible.

On all syllabuses, you are required to know in some detail the functioning of the external sense organs, the eye and ear.

The human eye

The human eye is a remarkable device for focusing light rays and converting them into nerve impulses. These impulses are carried along nerve fibres to the brain where an image of remarkable precision is produced.

You are unlikely to be asked to draw a detailed fully labelled diagram of the human eye (see Fig. 9.4). However, you may well be given a diagram which you must label. Alternatively, you may have to draw simplified diagrams to illustrate how the eye works or to illustrate how visual defects are caused and how they can be corrected (see Fig. 9.7a and 9.7b).

The eye functions by focusing an image on the light-sensitive layer, the retina, at the back of the eye. Light entering the eye takes the following path: conjunctiva → cornea → aqueous humour → pupil → lens → vitreous humour → retina.

To help you remember the structure of the eye, you can think of it as a number of layers. The conjunctiva (the outer covering of the front of the eye) is joined to the eyelids; the cornea and sclerotic layer form

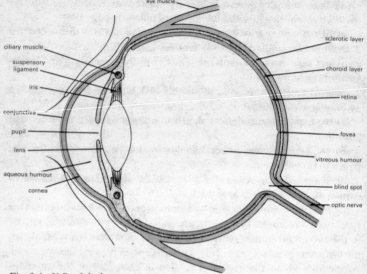

Fig. 9.4. *V.S. of the human eye*

a continuous ring; the choroid layer, iris, ciliary body and lens form another ring, while the retina forms the inner lining.

Here are the parts of the eye with a brief mention of their function:

Conjunctiva: a thin layer of transparent cells protecting the cornea.

Cornea: the transparent part of the sclerotic layer which refracts (bends) light rays towards the retina.

Pupil: the dark spot in the centre of the eye as seen from the front. It is a hole in the iris (below) through which all light enters the eye.

Iris: the coloured part of the eye. It is a muscular diaphragm which alters the size of the pupil and thus adjusts the amount of light entering the eye. It is made up of radial and circular muscles (Fig. 9.5).

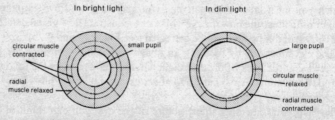

Fig. 9.5. *Altering the amount of light entering the eye: the iris/pupil response to light*

Aqueous humour: a clear watery fluid which supports the lens and provides the lens and cornea with food and oxygen.

Lens: made up of transparent living cells. It is elastic and focuses light rays on to the retina to form an image.

Ciliary muscle: circular muscles which alter the shape of the lens during focusing (accommodation).

Suspensory ligaments: which hold the lens in place and are attached to the ciliary muscles.

Vitreous humour: clear jelly-like substance which supports the eyeball.

Retina: contains the light-sensitive sensory cells, the rods and cones.

Fovea: the most sensitive part of the retina where most light rays are focused. It contains colour-sensitive cones only.

Blind spot: the point where the optic nerve leaves the eye and blood vessels enter and leave. There are no light-sensitive cells here.

Choroid layer: a dark layer which stops light being reflected. It contains blood vessels which supply food and oxygen to other parts of the eye.

Sclerotic layer: the tough outer covering of the eye. It maintains the shape of the eyeball.

Eye muscles: these are attached between the sclerotic layer and the skull. They move the eyeball.

Optic nerve: this contains the bundle of nerve fibres which carry nerve impulses to the brain.

Controlling the amount of light entering the eye

The pupil size is adjusted by reflex action according to the prevailing light conditions (light intensity). In bright light damage to the retina is prevented, and in dim light, maximum use is made of available light.

In bright light, the circular muscle of the iris diaphragm contracts and the pupil becomes smaller (see Fig. 9.5). This prevents too much light entering the eye. In dim light, the radial muscles contract, and the pupil becomes larger, allowing more light to enter (see Fig. 9.5).

Pupil size is also affected by focusing distance (see below).

Focusing (accommodation)

The eye is able to focus on objects at different distances by altering the shape (curvature) of the lens. When an object is a long distance away, the light rays reflected from it are almost parallel when they enter the eye. They have to be refracted (bent) by a small amount to be focused on the retina. A thin lens shape is required. The ciliary muscles relax and the pressure of the eye fluids (humours) pushes out and tightens the suspensory ligaments, and the lens is pulled out into a thin shape (see Fig. 9.6a).

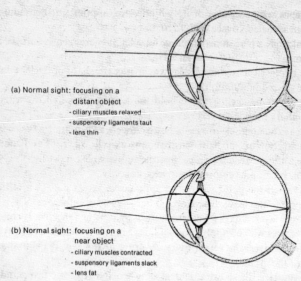

(a) Normal sight: focusing on a
       distant object
   - ciliary muscles relaxed
   - suspensory ligaments taut
   - lens thin

(b) Normal sight: focusing on a
       near object
   - ciliary muscles contracted
   - suspensory ligaments slack
   - lens fat

Fig. 9.6. *The human eye: focusing* (*accommodation*)

When an object is close to the eye, the light rays reflected from it are diverging (moving apart) when they enter the eye. They need to be refracted to a larger degree to be focused on the retina. A fatter lens shape is required. The ciliary muscles contract, the suspensory ligaments slacken, and the elastic lens springs back to a fatter shape (see Fig. 9.6b).

For technical reasons, the pupil size tends to be smaller when focusing on a close object.

Defective vision

Questions on sight defects, together with the means of correcting them, are popular.

1. **Short sight** (eyeball too long): a short-sighted person can focus on near objects, but far-away objects always appear out of focus. Because the eyeball is too long (from front to back) the point of clear focus is in front of the retina (see Fig. 9.7a). Short-sightedness can be corrected using spectacles (glasses) with a diverging lens; one that is thinner in the middle and fatter at the sides (see Fig. 9.7a).

2. **Long sight** (eyeball too short): a long-sighted person can focus on far-away objects but near objects always appear out of focus. Because the eyeball is too short from front to back the point of clear focus is behind the retina (see Fig. 9.7b). Long-sightedness can be corrected using

spectacles with a converging lens; one that is fatter in the middle and thinner at the edges (see Fig. 9.7b).

In old age, a person may become long-sighted, not because their eyeball

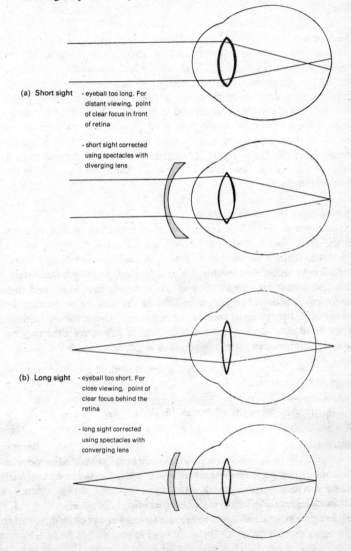

(a) Short sight   - eyeball too long. For distant viewing, point of clear focus in front of retina

  - short sight corrected using spectacles with diverging lens

(b) Long sight   - eyeball too short. For close viewing, point of clear focus behind the retina

  - long sight corrected using spectacles with converging lens

Fig. 9.7. *The human eye: sight defects and their correction*

is too long, but because their lens has become inflexible. The lens remains thin for distant vision and does not fatten for close-focusing. For close-up work older people usually require 'reading glasses' which have converging lenses.

The retina

The retina is made up of two types of sensory cell, **rods** which are sensitive to low light intensities, and **cones**, which respond only to bright light and detect colour. There are thought to be three types of cone cell in mammals, each responding to one of the primary colours, red, green and blue. Light of other colours and shades stimulates varying proportions of these cones. Impulses from these pass to the brain where the sensation of colour is produced. In man, cones are concentrated on the fovea, the most sensitive part of the retina. Nocturnal animals such as owls have many rods and few cones giving them better night vision but poor colour discrimination.

Stereoscopic (binocular) vision

Stereoscopic or three-dimensional vision occurs when both eyes focus on the same object. Each eye receives a slightly different view of the object and from these the brain forms a three-dimensional impression. Stereoscopic vision enables distance and depth to be judged accurately. Man and predatory animals such as cats and owls have well-developed stereoscopic vision. Their eyes are placed on the front of the head looking forward. Prey animals such as the rabbit have their eyes on the sides of the head so that they have a wide field of vision for detecting an approaching predator. They have little or no stereoscopic vision.

On certain syllabuses a knowledge of the compound eye of arthropods (page 305) is required.

Now try the following questions:

1. *Give labelled diagrams and explain the functions of*
(a) *ciliary muscles and suspensory ligaments of the eye in focusing.*

[LON]

2. (a) *At rest the human eye is focused on infinity. Explain what happens when the human eye focuses on a brightly-lit, nearby object.* [8]
(b) *Explain how spectacle lenses are used to*
(i) *enable a long-sighted person to see a nearby object clearly.* [2]
(ii) *enable a short-sighted person to see a faraway object clearly.* [2]

The human ear

The human ear is a sense organ with two quite separate functions: hearing and balance.

The structures concerned with hearing convert vibrations in the air (sound waves) into nerve impulses which are interpreted in the brain as sound. The structures concerned with balance detect changes in the orientation and movement of the head relative to gravity.

The ear is divided into the outer, middle and inner ear (see Fig. 9.8).

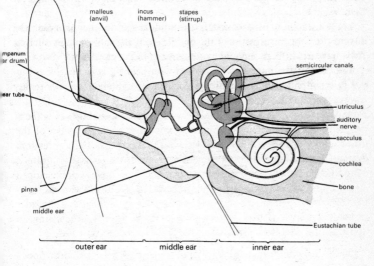

Fig. 9.8. *Diagram of the human ear*

Hearing involves all three parts, while balance involves only the inner ear.

Hearing

The **outer ear** consists of the pinna which collects sound waves and channels them down the ear tube to the tympanum (ear drum) which is caused to vibrate.

The **middle ear** is an air-filled cavity separated from the outer ear by the ear drum. Opposite the ear drum is the oval window. Vibrations of the ear drum (caused by sound waves) are transmitted to the oval window by three tiny bones, the **ear ossicles** (the malleus, incus and stapes). This arrangement magnifies the force of the vibrations about twenty times, the reason being to set up vibrations in the fluid of the inner ear. (NOTE. A greater force is required to set up vibrations in fluid than in air.)

The Eustachian tube connects the middle ear to the pharynx and ensures that the air pressure of the middle ear is the same as that outside. Any pressure difference could cause damage to the ear drum.

The part of the **inner ear** concerned with hearing is the **cochlea**. Fig. 9.9 is a simplified diagram of the cochlea straightened out (in reality

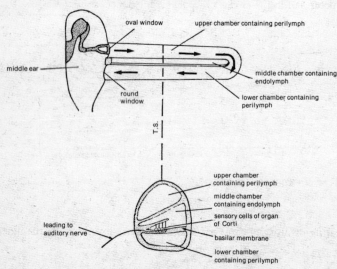

Fig. 9.9. *Detail of inner ear showing the cochlea straightened out*

it is coiled). When the oval window moves, vibrations are set up in the fluid called **perilymph**. These vibrations pass along the upper chamber and back to the round window along the lower chamber. As they do so, the **basilar membrane** is moved up and down. This causes the hairs of the sensory cells on the membrane to be pulled. It is here, at the **organ**

of Corti, that these sensory cells convert vibrations into nerve impulses.

The basilar membrane becomes thicker further away from the two windows and it is thought that the membrane here is sensitive to lower pitch (lower frequency) sounds. Nearer the windows the membrane is sensitive to higher pitched (higher frequency) sounds. The nerve impulses from the sensory hair cells are sent along the **auditory nerve** to the brain where they are interpreted. Clearly the hearing mechanism must be extremely sensitive. It can detect differences in pitch and loudness and also in the quality of sound. A person can distinguish, for example, different musical instruments playing at the same time.

Q. (*a*) *Make a labelled diagram to show the structure of the organs of hearing in a mammal.* [12]
(*b*) *Explain the mechanism of hearing in a mammal.* [8]

[OXF]

Balance

The **utriculus, sacculus** and the three **semicircular canals** are the structures of the inner ear concerned with balance (see Fig. 9.10).

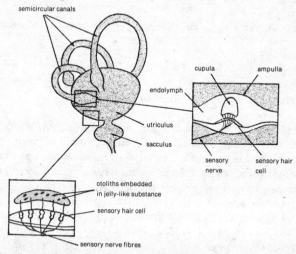

semicircular canals

cupula   ampulla

endolymph

utriculus

sacculus

sensory nerve   sensory hair cell

otoliths embedded in jelly-like substance

sensory hair cell

sensory nerve fibres

Fig. 9.10. *Detail of inner ear showing structures concerned with balance*

The utriculus and sacculus detect the position of the head relative to the pull of gravity. Within these structures are sensory hair cells, the hairs of which are embedded in a jelly-like substance containing granules of chalk called **otoliths**. Any change in posture tends to displace the otoliths and alter the pull on the hairs of the sensory cells; this stimulates the nerve fibres and impulses are transmitted to the brain giving the new position of the head relative to gravity.

The semicircular canals detect changes in the direction of movement of the head. The three canals are set at right angles to one another and contain the fluid called **endolymph**. At one end of each canal is a swelling, the **ampulla**, which contains sensory hair cells with their hairs embedded in a cone of jelly called the **cupula**. When the head is moved the fluid in the canal 'lags behind' and causes the cupula to be bent over. Which cupula is displaced will depend on which direction the head is moving. The brain receives nerve impulses from the various sensory hair cells in the three ampullae and can determine the direction of movement of the head.

The brain is able to achieve balance, posture and muscular coordination in the body, not only using the information from the sensory hair cells of the inner ear, but also from information received from the eyes, and proprioreceptors in the muscles, joints and tendons.

Q. (*a*) *Make labelled diagrams to show the structure of those organs in the ear of a mammal that perceive gravity and explain how they function.*

[9]

(*b*) *In what other ways do mammals perceive gravity and movement?*

[3]

[OXF]

9.5. *The Mammalian Endocrine System*

The **endocrine system** is made up of **endocrine (ductless) glands** which secrete their products, **hormones**, in minute quantities directly into the blood, rather than through tubes (ducts). These hormones travel through the bloodstream and exert an effect on **target** organs, tissues or cells, by speeding up or slowing down their activity.

The endocrine system, sometimes working in conjunction with the nervous system, controls and coordinates growth, development and activity within the organism. The nervous system and endocrine system are similar in two ways:

1. They are both set into action by a stimulus to produce a response.

2. They both involve chemical transmission. In the case of the nervous system this occurs only across synapses, whereas in the endocrine system the chemical, a hormone, travels considerably further.

More usually, the *differences* between nervous and hormonal action are stressed. These are summarized in Table 9.1.

Table 9.1 *Differences between nervous and hormonal action*

	Nervous	Hormonal
1. Method of transmission	Electrical impulse passing along nerve fibre (chemical transmission across synapses)	Chemical passing through bloodstream
2. Speed of transmission	Rapid	Slow
3. Speed of response	Immediate	Usually slow, e.g. thyroxine, but may be fast, e.g. adrenalin
4. Duration of response	Short-lived	Longer-lasting
5. Localization of response	Response very localized	Response usually widespread, e.g. adrenalin

Questions asking you to compare nervous and hormonal action are common. For example:

Q. Give three differences between the transmission of information by nerve impulses and the transmission of information by hormones. [6]

[OXF]

The main endocrine glands in the human body are the pituitary, thyroid, adrenals, pancreas and the reproductive organs. The positions of these glands are shown in Fig. 9.11. The pituitary gland, sometimes

Fig. 9.11. *The main endocrine glands in the human body*

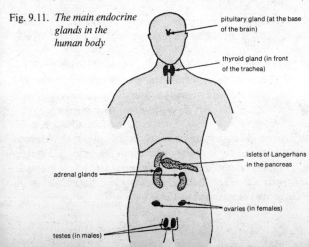

pituitary gland (at the base of the brain)

thyroid gland (in front of the trachea)

islets of Langerhans in the pancreas

adrenal glands

ovaries (in females)

testes (in males)

called the 'master gland', coordinates and controls the activity of the other glands. It releases hormones which stimulate these glands to release their own hormones.

The various endocrine glands are listed in Table 9.2 together with their hormones, the stimulus which triggers release of the hormone, and the effect of the hormone on target tissue. Knowledge of these glands, their hormones and their effects is frequently demanded in multiple-choice or short-answer papers.

Table 9.2. *Summary of endocrine glands and hormones in human*

Endocrine gland	Hormone	Stimulus	Effect
Pituitary gland (master gland)	Anti-diuretic hormone (ADH)	High blood O.P.	Increases reabsorption of water from the kidney tubule (see page 172)
	Growth hormone	?	Stimulates cell division and protein synthesis
	Thyroid-stimulating hormone (TSH)	Low level of thyroxine (feedback)	Stimulates release of thyroxine by thyroid gland
	Follicle-stimulating hormone (FSH)		
	Luteinizing hormone (LH)	See page 257	
	Oxytocin		
	Prolactin		
Thyroid gland	Thyroxine	TSH (see above)	Increases body's metabolic rate
Islets of Langerhans in the pancreas	Insulin	High blood glucose level	Lowers blood glucose level, stimulates glucose uptake by cells

Endocrine gland	Hormone	Stimulus	Effect
Adrenal gland			
Medulla	Adrenalin (fight or flight hormone)	Danger, fright, stress	Increases heart beat. Increases breathing rate. Diverts blood flow from gut to brain and muscles. Raises blood glucose level
Cortex	Aldosterone	Low blood O.P.	Increases reabsorption of mineral salts from the kidney tubule
Ovary (females only)	Oestrogen	See page 257	Stimulates development and maintenance of secondary sexual characteristics (see also page 258)
	Progesterone	See page 257	See page 257
Testis (males only)	Testosterone	See page 258	Stimulates development and maintenance of secondary sexual characteristics (see also page 259)

The level of circulating hormone is controlled by a **'feedback'** process; a hormone adjusts its own output by affecting the endocrine glands that cause its secretion.

The brain monitors the level of hormone in the blood. If too much is present the pituitary gland is blocked from releasing the appropriate stimulating hormone; if too little is present, more stimulating hormone is released. Feedback control provides an effective means of monitoring and maintaining suitable hormone levels in the body. Two examples are:

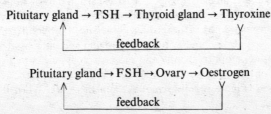

Pituitary gland → TSH → Thyroid gland → Thyroxine

feedback

Pituitary gland → FSH → Ovary → Oestrogen

feedback

In the human female menstrual cycle (page 257), feedback between several hormones generates cyclical changes. Several hormones often affect the level of a substance or an activity in the body. For example, blood glucose level is controlled by three hormones: insulin, thyroxine and adrenalin.

9.6. *Homeostatic Mechanisms*

In Chapter 8 we looked at the concept of homeostasis – keeping conditions within the body constant. The mechanisms involved in homeostasis all follow the same basic plan and involve either nervous or hormonal action, or a combination of the two. This basic plan can be represented as follows:

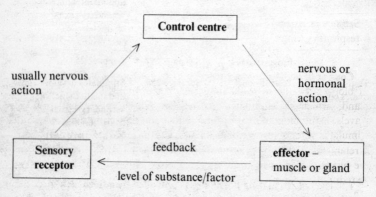

As with the control of hormone levels in the blood, the level of other substances and factors in homeostasis is controlled by feedback.

To understand the above scheme it is best to apply it to a few examples. Once understood, this scheme is very useful in checking your knowledge and understanding of different homeostatic mechanisms. It is also a convenient way of remembering them.

1. **Control of breathing:** keeping the level of carbon dioxide in the blood constant. If the level of carbon dioxide in the blood rises this is sensed by the respiratory centre in the medulla oblongata of the brain (page 187). This sends out nerve impulses to the diaphragm and intercostal muscles which contract more frequently and more strongly, increasing the rate and depth of breathing (see page 126). As a result, excess carbon dioxide in the blood is more rapidly removed from the body. When the level

of carbon dioxide returns to normal, this is sensed by the respiratory centre. It sends out fewer nerve impulses to the diaphragm and intercostal muscles, and as a result the rate and depth of breathing return to normal.

The respiratory centre is the sensory receptor and control centre, while the diaphragm and intercostal muscles are the effectors:

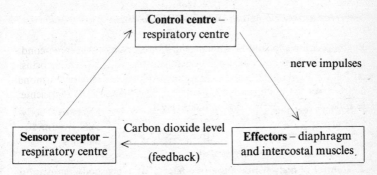

2. **Control of blood osmotic potential** (O.P.; see page 58): when the O.P. of blood rises, the hypothalamus sens nerve impulses to the pituitary gland, which is stimulated to release ADH (page 200). ADH travels in the blood to the kidney tubule and collecting duct, where it stimulates water reabsorption from the urine (page 172). As a result, water is retained in the body, thus tending to keep the blood O.P. down. When the blood O.P. returns to normal, feedback causes ADH secretion to be reduced.

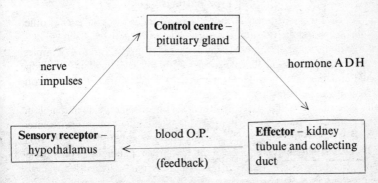

3. **Control of blood glucose level** (see page 94): when blood glucose levels are high, the hormone insulin is released by the pancreas. This stimulates the uptake of glucose by cells in the body, and the conversion of glucose to glycogen by the liver and muscles.

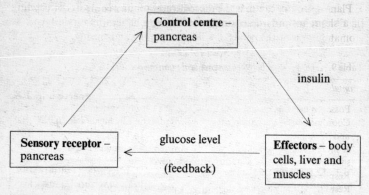

4. **Control of body temperature** (see page 175): when body temperature rises above normal this is sensed by the hypothalamus which sends nerve impulses to various effectors in the skin which give responses to increase heat loss and thus lower body temperature (see page 176). When the body temperature drops below normal, the hypothalamus sends nerve impulses to effectors in the skin which respond by reducing heat loss. At the same time skeletal muscles may be stimulated to contract (shivering).

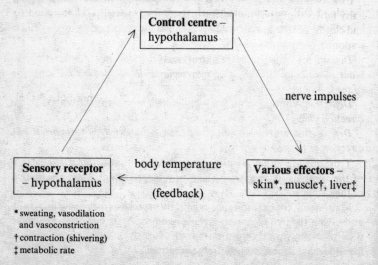

* sweating, vasodilation
  and vasoconstriction
† contraction (shivering)
‡ metabolic rate

9.7. *Plant Responses*

Plants are able to respond to light, gravity, temperature, chemical substances, water and, in some cases (such as climbing plants), to touch.

Plant responses are due to different rates of growth, as in the bending of a shoot towards light or, less often, to changes in turgidity, as in stomatal pores or the petals of a flower opening and closing.

Table 9.3. *Differences between animal and plant responses*

Animals	Plants
1. Possess a nervous system	Lack a nervous system
2. Coordination by nerve impulses and hormones	Coordination by hormones only
3. A short stimulus is usually sufficient	A prolonged stimulus is usually required
4. Responses are usually rapid	Responses are usually slow
5. Responses usually involve movement	Responses usually involve growth
6. The effect is usually temporary	The effect is often permanent

Most plant responses can be classed as either **nastic responses** or **tropic responses**.

Definition: Nastic responses *are movements by part of the plant in which the direction of response is not determined by the direction of the stimulus.*

Examples of nastic responses are the opening of the petals of the daisy flower during daylight and their closure at night, and the opening and closing of stomata with changes in light intensity (see page 155). Such responses are non-directional.

Definition: Tropic responses (tropisms) *are growth movements by the plant in which the direction of movement is determined by the direction of the stimulus.*

Definition: Phototropism *is a plant growth movement in response to the direction of light.*

Definition: Geotropism *is a plant growth movement in response to the direction of the pull of gravity.*

Tropisms are either **positive**, when movement is towards the source of the stimulus, or **negative**, when movement is away from the source of the stimulus.

Main stems, for example, are **negatively geotropic** (they grow upwards, away from the pull of gravity), while main roots are **positively geotropic** (they grow downwards, towards the pull of gravity). A plant stem growing towards a light source is **positively phototropic**.

Tropisms are survival responses. For example:
1. They enable seedlings to become established with the roots in the soil and leaves in the air.
2. They enable a plant to change its growth pattern at any time in its life to suit changing circumstances. For example, if a plant survives being blown over in the wind, new growth will make the stem curve upwards under the influence of negative geotropism.

Tropisms may be investigated experimentally using seedlings, since these are particularly sensitive to the stimuli of light and gravity. The **coleoptiles** of oat or maize are traditionally used because they do not have leaves or buds to confuse findings.

Phototropism experiments

A simple experiment to determine the effect of light on the growth of oat seedlings is shown in Fig. 9.12. As far as possible, conditions are the same in all three pots, except for the presence, absence or direction of light.

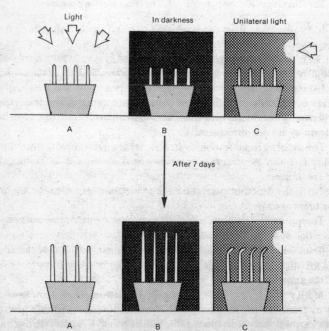

Fig. 9.12. *Experiment to determine the effect of light on oat seedlings*

The seedlings in C receive unilateral light (light from one side) and show a growth curvature towards the light. The seedlings in A show normal growth. The seedlings in B, without light, grow tall and spindly. They are yellow and lacking in chlorophyll. This type of response to lack of light is termed **etiolation**. It is a good example of a survival response. In the wild, if the seed lands in dense undergrowth where it is shaded from light, it will grow rapidly upwards to reach the light, where it can start to photosynthesize and manufacture its own food. Without light, the seedling cannot photosynthesize and will eventually die.

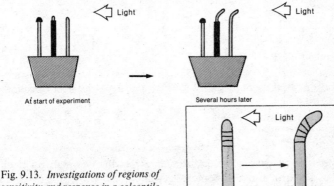

Fig. 9.13. *Investigations of regions of sensitivity and response in a coleoptile*

The experiment shown in Fig. 9.13 can be used to investigate the region of sensitivity and region of response in the phototropic response of a coleoptile. The results indicate that:

1. The shoot tip is sensitive to light. The phototropic response only occurs in those shoot tips which are exposed to light.
2. The region of growth response (bending) occurs immediately behind the tip in the region of elongation. This can be better shown by marking this region with lines (see inset, Fig. 9.13).
3. There must be some 'influence' passing down from the tip to the region of cell elongation. The experiment does *not* indicate what this influence is (be careful not to draw conclusions not warranted by the results).

Further experiments can be conducted to investigate this influence. In the experiment in Fig. 9.14 the tip of a coleoptile is removed and placed for several hours on an agar block. The block is then placed on the coleoptile stump. For the control, the process is repeated but the coleoptile tip discarded and an untreated agar block placed on the stump. The results indicate that some substance present in the coleoptile tip,

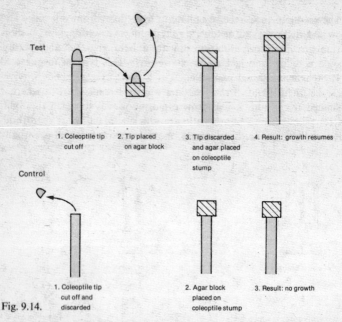

Test

1. Coleoptile tip cut off   2. Tip placed on agar block   3. Tip discarded and agar placed on coleoptile stump   4. Result: growth resumes

Control

1. Coleoptile tip cut off and discarded   2. Agar block placed on coleoptile stump   3. Result: no growth

Fig. 9.14.

and transferred to the agar block, stimulates growth. A normal agar block used as a control does not promote growth.

The above experiment indicates the presence of a growth-stimulating substance in the coleoptile tip. The next experiment (Fig. 9.15) indicates

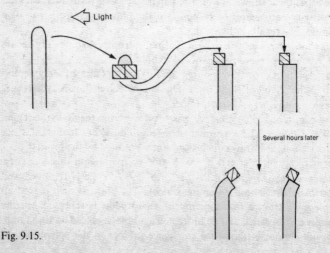

Light

Several hours later

Fig. 9.15.

that such a substance is responsible for growth curvatures in response to light. The agar block from the side of the shoot furthest from the light causes a greater curvature than the block from the side nearest the light. The generally accepted conclusion is that light somehow causes more of a growth-stimulating chemical to gather on the side of the coleoptile tip furthest from the light, i.e. on the shaded side.

There is now strong evidence that in many plants a chemical called **auxin** is produced by the region of actively dividing cells at the tip of the shoot (the apical meristem) (see Fig. 9.16). More auxin passes down

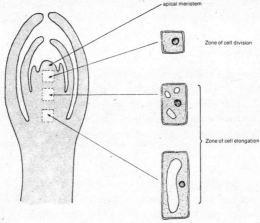

Fig. 9.16. *Shoot tip*

the shaded side of the shoot where it stimulates growth of cells in the region of cell elongation, causing them to elongate. The result is a growth curvature towards the light (see Fig. 9.17).

Fig. 9.17. *Cell elongation causing a growth curvative towards the light*

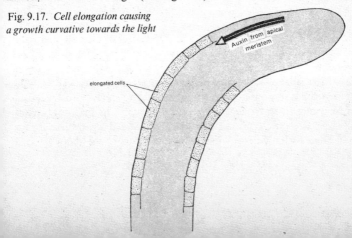

The general name auxin is given to those growth substances which are produced at the apices (tips) of shoots and roots. One auxin which has been isolated from plants is indoleacetic acid (IAA). Auxins are the most common of a range of plant hormones which control plant responses. Other hormones influence not only growth but, for example, flowering, bud sprouting, leaf shedding and seed dormancy.

Synthetic compounds based on natural auxins are used by farmers, market gardeners and horticulturalists to increase the yield of crops. Compounds related to IAA are sprayed on lawns as selective weed killers; broad-leaved plants (weeds) absorb large quantities, 'outgrow' themselves and die, while growth in the narrow-leaved grasses is stimulated.

Geotropism experiments

Roots and shoots show growth movements in response to the pull of gravity. When a young bean seedling is placed in a horizontal position in the dark, and well supplied with water, growth of the shoot curves upwards and root growth curves downwards (see Fig. 9.18). The shoot is negatively geotropic and the root is positively geotropic.

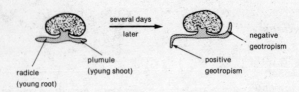

Fig. 9.18. *Geotropism in a germinating bean*

In performing experiments to examine the effect of the pull of gravity, a control is established using a **clinostat** (a slowly rotating drum). This rotates the seedling, subjecting it to the pull of gravity equally on all sides. The results of a typical experiment are shown in Fig. 9.19.

In geotropic responses, auxins are thought to migrate to the lower side of root and shoot under the pull of gravity. In the root, auxins inhibit cell elongation, thus causing the root to bend downwards. In the shoot, auxins stimulate cell elongation, thus causing the shoot to bend upwards.

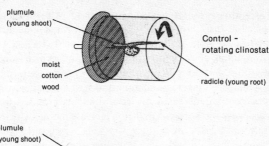

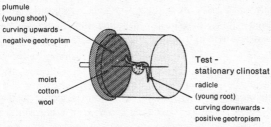

Fig. 9:19. *Investigation of geotropism in a germinating bean*

Definitions

Sensitivity (irritability)
Reflex action
Nastic response
Tropic response
Phototropism
Geotropism

Key words

Coordination
Nervous system
Nerve impulses
Endocrine system
Hormones
Central nervous system
Peripheral nervous
  system

Autonomic nervous
  system
Cerebellum
Thalamus
Cerebral cortex
Motor areas
Sensory areas
Association areas

Semicircular canals
ADH
Growth hormone
TSH
FSH
LH
Oxytocin
Prolactin

Sensory receptors
Effectors
Sensory neurones
Relay (intermediate) neurones
Motor neurones
Spinal reflexes
Conditioned reflexes
Spinal cord
White matter
Grey matter
Central canal
Brain
Cerebrum
Medulla oblongata

Hypothalamus
External receptors
Skin receptors
Olfactory organs
Taste receptors
Taste buds
Light receptors
Retina
Organ of Corti
Sensory hair cells
Internal receptors
Respiratory centre
Proprioreceptors
Utriculus
Sacculus

Thyroxine
Insulin
Adrenalin
Aldosterone
Oestrogen
Progesterone
Testosterone
Feedback
Coleoptile
Etiolation
Agar
Meristem
Cell elongation
Auxin
Indoleacetic acid (IAA)

Exam Questions

1. Irritability in living things may be defined as an organism's ability to respond to a stimulus. Most instances of irritability fit into the following scheme.

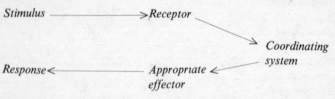

   Two instances of irritability are a person dropping a hot object and the phototropic curvature of a plant shoot.
(i) Explain fully how each of the examples given fits into this overall scheme for irritability. [12]
(ii) Name three *other stimuli to which humans respond and for each state the appropriate receptor.* [3]
(iii) What is the value to the plant of the phototropic response of its shoot? [3]
(iv) The flowers of some plants close at night and open during the day. Suggest two *stimuli which would cause this response.* [2]
(v) How does the nature of the 'closing' response differ from a phototropic response? [2]

[AEB]

2. (*a*) What is a reflex action? [2]

(*b*) Describe, using a simple diagram, a reflex arc for a named *reflex action*. Include in your diagram the specific receptor, the effector, and the types of neurone involved. [8]

(*c*) In what way is the reflex action you describe above of value to the person concerned? [1]

3. (*a*) Make a labelled diagram to show the structure of the spinal cord of a mammal as seen in transverse section. [9]

(*b*) Indicate on your diagram the position of neurones that take part in the reflex arc and show the direction in which impulses travel. [5]

(*c*) Mark on your diagram with an X the point at which it would be practical to make a cut that would prevent pain from being felt. [1]

Mark on your diagram with an O the point at which it would be practical to make a cut that would prevent a response, but allow pain to be felt.

[1]

[OXF]

4. For a named directional stimulus state how a flowering plant may:

(*i*) perceive it;

(*ii*) respond to it. [7]

[OXF]

10. Support and Movement

All organisms have to support themselves, to a greater or lesser degree, in order to maintain body shape. Self-support is usually achieved using some kind of skeleton. The degree to which an organism must support itself is largely determined by the environment it lives in. In aquatic environments, the water provides a high degree of support and less self-support is required. Many aquatic invertebrates, *Hydra* (page 328) for example, do not possess a skeleton.

On land, the air provides little support and much more self-support is required. Land vertebrates depend almost entirely on their skeleton for support; it raises the body off the ground and makes locomotion (movement from one place to another) possible. In addition, many of the internal organs are suspended from the skeleton. Flowering plants on land are supported by the turgor of packing tissues (page 60) and by rigid supporting tissues such as xylem vessels (page 149).

Movement is a characteristic feature of living organisms, but is much more obvious in animals than in plants. Animals, in general, can move more rapidly and most show locomotion. Only a few simple plants, e.g. *Chlamydomonas* (page 310), show locomotion, and most plants move only part of the body, and then only slowly (see page 205).

The difference in locomotory ability between animals and plants is related to their method of nutrition. On land, green plants need simply to bask in the sun and obtain water and minerals from the soil in order to feed (see page 64). Animals, on the other hand, are heterotrophic (page 63) and require ready-made organic food. They may have to travel long distances to find a suitable source. Being freely mobile has other advantages: they can avoid predators, seek out mates, and disperse to new areas.

10.1. *Types of Skeleton*

With only a few exceptions, **locomotion** is brought about by contraction of muscles acting against some kind of skeleton. In some simpler organisms, e.g. *Amoeba* (page 326), locomotion does not involve a skeleton in the normal sense.

All **skeletons** are resistant to compression, and provide a rigid framework for the body. Three types of skeleton are found in animals and they are often important in protection as well as support and locomotion.

1. A **hydrostatic** skeleton is found in the earthworm (page 333). It is a fluid skeleton and as such is incompressible (does not change in volume when a pressure is applied), but it does change shape. Muscles contract against the fluid skeleton and cause changes in body shape which give rise to locomotion. The pressure of fluid pressing out against the muscles supports the body.

2. An **exoskeleton** (outer skeleton) is found in arthropods, e.g. insects (page 335). This rigid outer covering is called a **cuticle**. At the joints the skeleton is soft and flexible to allow for movement. Limbs are jointed and hollow and have inward projections for muscle attachment. To allow for growth, the exoskeleton must be periodically shed (ecdysis) and a new one laid down.

3. An **endoskeleton** (internal skeleton) is found in vertebrates. It is made of hard material, either **bone** or **cartilage**. The limbs are jointed, and muscles are attached to the bones, which act as levers.

10.2. *The Mammalian Skeleton*

The skeleton of mammals is composed largely of bone. Bone is not dead material, but is made up of a non-living part, mostly calcium phosphate, together with the protein collagen, secreted by living bone cells. The cells are found in spaces within the non-living part, and they receive food and oxygen from blood vessels which run through the bone. Apart from its role in support and movement, bone is the site where red and white blood cells and platelets are made (in the bone marrow, see page 134). Bone is also the site of calcium storage. If the body tissues are short of calcium, bone cells release some calcium salts into the blood. They are replaced when the body's intake of calcium is resumed.

Cartilage is found closely associated with bone. It is softer than bone and is found at joints where one bone articulates with (moves against) another. It is a smooth material and produces very little friction at points of contact. Its ability to 'give' slightly when compressed means it has a cushioning effect and absorbs shocks.

Associated with the bones of the skeleton are **ligaments** (which attach bone to bone), **tendons** (which attach muscle to bone) and **muscles**, which when they contract move one bone relative to another.

The mammalian skeleton has two parts (see Fig. 10.1):

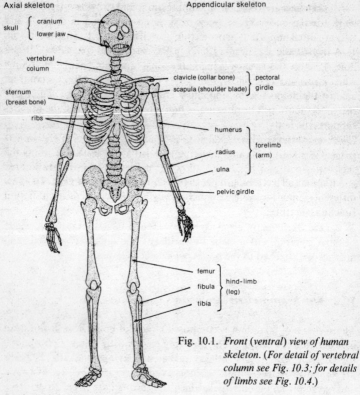

Fig. 10.1. *Front* (ventral) *view of human skeleton. (For detail of vertebral column see Fig. 10.3; for details of limbs see Fig. 10.4.)*

1. the axial skeleton
2. the appendicular (hanging) skeleton.

1. The **axial** skeleton consists of the skull, vertebral column (backbone), ribs and sternum (breast bone).

The **skull** consists of the cranium (brain box) and jaws. The cranium is made of a number of fused bones which enclose and protect the brain and organs of hearing, sight and smell. The upper jaw is fixed to the cranium, while the lower jaw is jointed to it.

The **vertebral column** in man is made up of 33 bones called **vertebrae** (singular: vertebra). The vertebral column performs three main functions:
(a) It protects the spinal cord (see page 185).
(b) It enables flexible movement of the body.
(c) It determines the body's posture.

Each vertebra is separated from the next by a disc of cartilage which acts as a shock absorber and allows limited movement.

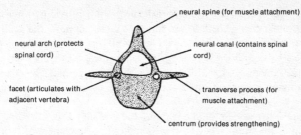

Fig. 10.2. *Generalized structure of a vertebra*

Although all vertebrae have the same basic structure (Fig. 10.2), in different regions of the backbone they are specialized for different functions (see Fig. 10.3):

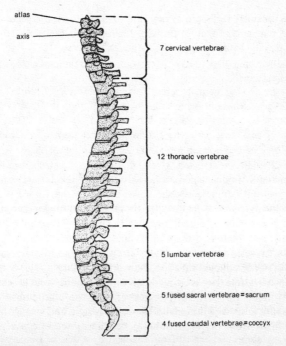

Fig. 10.3. *Side (lateral) view of the human vertebral column*

(i) The 7 **cervical** vertebrae support the neck while providing some flexibility of movement. The first vertebra, the **atlas**, supports the skull and allows nodding ('yes') movements. The second vertebra, the **axis**, projects into the atlas and allows swivelling ('no') movements. The other vertebrae have small processes and also permit twisting of the neck. The vertebrae in this region have large holes parallel to the neural canal which allow blood vessels through to reach the brain.

(ii) The 12 **thoracic** vertebrae support the ribs and allow bending and rotation of the trunk. They have long spines for attachment of shoulder and back muscles. The transverse processes are reduced because of the presence of ribs.

(iii) The 5 **lumbar** vertebrae are thick-set with large processes and wide spines for attachment of muscles. These vertebrae provide the major support for the trunk and are subject to great stress. They allow bending and rotation of the trunk.

(iv) The 5 **sacral** vertebrae are fused together for strength and form part of the pelvis. They transmit the weight of the body to the pelvis and hips.

(v) The 4 **caudal** vertebrae in man are fused to form the **coccyx**. They have no particular function in man. In the rabbit there are 16 caudal vertebrae; they form the support for the tail.

The mnemonic on pages 19–20 can help you to remember the different types of vertebra.

The **ribs** and **sternum**, together with the thoracic vertebrae, enclose the thoracic cavity and protect the heart and lungs. Intercostal muscles are attached to the ribs. When they contract and relax they alter the volume of the thoracic cavity producing ventilation of the lungs (see page 124). The ribs are connected to the sternum by cartilage.

2. **The appendicular skeleton** (girdles and limbs): the **girdles** are box-like arrangements of bone which connect the limbs to the axial skeleton.

The **pectoral** (shoulder) **girdle** is connected to the vertebral column by muscles and tendons. The girdle consists of a pair of **scapulas** (shoulder blades) and a pair of **clavicles** (collar bones). The forelimbs (in man, the arms) are jointed to the scapula and are attached to the pectoral girdle by muscles and tendons. This arrangement allows great mobility of the forelimbs coupled with shock-absorbing ability.

The **pelvic girdle** (pelvis) is made up of fused bones which in turn are fused to the sacral vertebrae (see above). The two hind-limbs (in man, the legs) are jointed to the pelvis. The pelvis is rigid and strong to enable the upthrust from the hind-limbs to be transmitted effectively to the vertebral column.

The **limbs** are based on the pentadactyl ('five finger') plan typical of

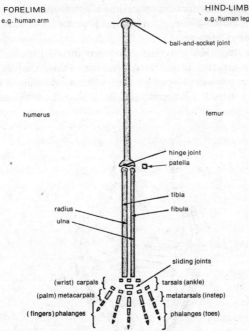

FORELIMB
e.g. human arm

HIND-LIMB
e.g. human leg

ball-and-socket joint

humerus

femur

hinge joint
patella

tibia

radius
fibula

ulna

sliding joints

(wrist) carpals { tarsals (ankle)

(palm) metacarpals { metatarsals (instep)

(fingers) phalanges { phalanges (toes)

Fig. 10.4. *Basic plan of the limb skeleton (the pentadactyl limb)*

higher vertebrates (see Fig. 10.4). The basic plan is modified according to:

(a) whether it is a fore- or hind-limb
(b) the method of locomotion, e.g. swimming, walking or flying.

Having looked at the mammalian skeleton try the following question:

Q. (a) Give three functions of the skeleton of a mammal. [3]
(b) What are the major chemical constituents of bone? [2]
(c) Name the principal regions of the spine and for each region give:
(i) one special function;
(ii) one adaptation to performing that function. [15]

[OXF]

Joints

A joint is the place where two bones meet. Joints can be broadly classified as movable, slightly movable or immovable.

1. **Immovable** joints (sutures) join bones together in the cranium and pelvic girdle. Where bones meet they are connected by the fibrous protein **collagen**.

2. **Slightly movable** (cartilaginous) joints are found where bones are separated by cartilage and tightly bound together by ligaments, e.g. the joints between vertebrae.

3. **Movable** (synovial) joints are associated with the limbs. Where the bones come together they are covered with shiny, slippery cartilage. Between the cartilages is **synovial** fluid which allows almost frictionless movement between the bones. The fluid is secreted by the synovial

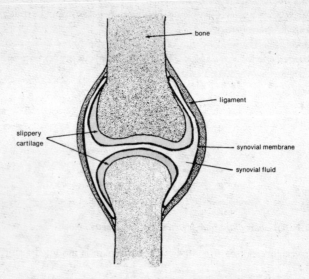

Fig. 10.5. *Structure of a synovial joint*

membrane which surrounds the whole joint (see Fig. 10.5). Ligaments hold the bones together. The arrangement shown in Fig. 10.5 applies to the ball-and-socket and hinge joints described below.

Movable joints fall into one of three groups:

(a) **Ball-and-socket joints** (also called universal joints): these allow movement in all planes; up and down, side to side, and rotation. These joints are found between the pelvic girdle and femur (hipbone) and between the scapula (shoulder blade) and humerus (upper arm).

(b) **Hinge joints:** these allow movement in one plane only, like the hinge of a door. They are found at the knee and elbow.

(c) **Sliding joints:** these are found where the flat surface of one bone glides across the surface of another. Such joints give flexibility to the wrist and ankle.

10.3. *Muscle*

Bones are moved relative to each other by contraction of muscles. There are three types of muscle, but only one type is associated with the skeleton:

1. **Striped (skeletal) muscle:**

(a) It is attached to the skeleton and provides the force required for body movement and locomotion.

(b) It contracts powerfully when stimulated by nerve impulses (page 182) but soon fatigues (tires).

(c) Under the microscope the elongated muscle fibres have a striped appearance (hence the name). There is no clear-cut separation between one cell and the next (see Fig. 10.6).

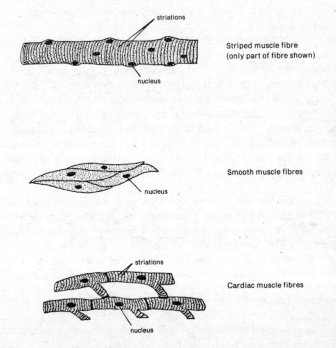

Fig. 10.6. *Types of muscle fibre*

(d) Muscle contraction is normally under conscious control, hence these muscles are sometimes called 'voluntary' muscles.

The other two types of muscle are:

2. **Smooth (visceral) muscle:**

(a) This is found in the walls of internal organs such as the gut and bladder.

(b) When stimulated by nerve impulses, it contracts less powerfully than striped muscle, but it is longer acting, i.e. it does not fatigue so easily.

(c) The cells are unstriped and are clearly separate from each other (Fig. 10.6).

(d) Muscle contraction is not under conscious control, hence these muscles are sometimes called 'involuntary' muscles.

3. **Cardiac (heart) muscle:**

(a) This is found in the walls of the heart.

(b) It is **myogenic** (contracts of its own accord *without* nervous stimulation). NOTE. The heart is under nervous control, but this is to speed up or slow down heartbeat, not initiate it (see page 143).

(c) Cardiac muscle is unfatiguing (if your heart stopped beating you would soon die).

(d) The cells are striped, individual cells are distinguishable and the cells (fibres) are branched (Fig. 10.6).

Muscles are made up of thousands of **muscle fibres** which have the ability to contract (shorten) and relax. When they contract, they exert a force. In the case of skeletal muscle this force usually moves one bone relative to another. When skeletal muscle relaxes it does not return to its original length of its own accord. It must be pulled out by gravity or, more usually, by another muscle contracting against it and pulling it out. In this way, skeletal muscles usually act against each other in **antagonistic pairs** (see Fig. 10.7). When one muscle contracts the other relaxes.

Muscle contraction requires energy. This is obtained in the form of ATP, produced by respiration (see page 110). Muscles therefore require a good blood supply delivering food and oxygen for aerobic respiration (page 107). This blood supply can be adjusted according to muscle need. At times of peak energy demand, e.g. during heavy exercise, aerobic respiration is supplemented by anaerobic respiration (see page 108).

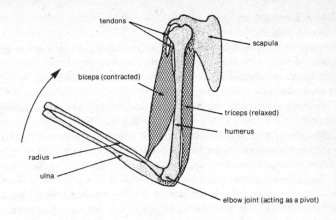

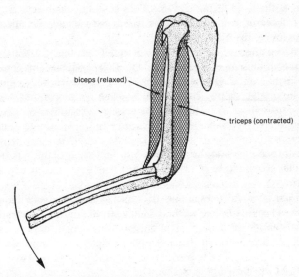

Fig. 10.7. *Movement at the elbow joint*

10.4. *A Limb Joint*

For most examination boards you need to be familiar with one *named* limb joint; how it works and how bones, muscles, tendons and nerves bring about movement at the joint. We will choose the elbow joint as our example.

The elbow joint is of the hinge type and permits and up and down movement of the lower arm. However, unlike the hinge joint at the knee, it also allows some rotation (you can twist your forearm).

The muscle which contracts and bends (flexes) a limb joint is called a **flexor muscle**; in this case it is the **biceps** (see Fig. 10.7). The muscle on the opposite side of the limb joint, and which straightens (extends) the limb, is called the **extensor muscle**, in this case the **triceps**. Notice the biceps muscle is larger than the triceps. This is because it is the major load-bearing muscle and acts against the pull of gravity. The triceps, on the other hand, is acting in the direction of the pull of gravity. With the palm of the hand facing upwards, contraction of the biceps muscle raises the forearm; contraction of the triceps muscle and relaxation of the biceps lowers the forearm.

To produce effective movement it is essential that the contraction of various sets of muscles is coordinated so that, for example, antagonistic muscles do not contract at the same time. Each muscle has both a motor and sensory nerve supply. Nerve impulses from the motor area of the cerebral cortex (page 187) stimulate the muscle to contract. Within the muscle are stretch receptors which, when stretched, send nerve impulses along sensory neurones to the spinal cord and brain. These receptors supply information about the position of the limbs and enable a pattern of muscular activity to be computed by the brain, so producing effective movement.

Muscles that move bones are usually attached in such a way that the bones act as levers with low mechanical advantage. Fig. 10.7 makes this clear. The muscles of the upper arm only contract over a short distance but, being attached near to the joint, they magnify the movement at the end of the limb. The biceps muscle may contract only 8 or 9 cm but the hand will move about 60 cm.

The skeletal muscles of the body are rarely in the state where one muscle of an antagonistic pair is completely relaxed. Even when the body is at rest, a proportion of the fibres in the antagonistic muscles remain in a state of contraction producing a **muscular tone** which holds the body in position.

Using your course notes to help you, answer the following two questions:

Q. Make a labelled diagram of a named *limb of a mammal to show the arrangement of the bones, tendons, muscles and nerves which bring about movement at a joint. Include details of the joint itself.* [11]

[OXF)

Q. Explain how nerves, muscles and joints are involved in the movement of the limb of a mammal. [25]

[LON]

10.5. *Types of Locomotion*

On certain syllabuses you are required to know a range of methods of locomotion in vertebrates, including swimming, crawling, leaping, walking, running and flapping flight. Check your syllabus requirements. We will examine two methods of locomotion which are common to most syllabuses: swimming and flapping flight.

Swimming in a bony fish, e.g. herring

The body of the fish is well adapted for movement in water:

1. The streamlined body form, backward-pointing scales and mucus covering provide little resistance to movement through water.
2. Various **fins** are present which provide
(a) propulsion
(b) stability.
3. The flexible body and vertebral column allows swimming movements.
4. Muscle blocks (myotomes) on either side of the vertebral column provide the power for swimming.
5. An air bladder may be present which provides buoyancy and makes the fish effectively 'weightless' in water.

Propulsion: the muscular tail has a large surface area which provides the force which propels the fish forward. This is achieved by alternate contraction and relaxation of muscle blocks (myotomes) lying on opposite sides of the vertebral column (Fig. 10.8b). The tail and body exert a sideways and backwards thrust against the water and the resistance provided by the water pushes the fish sideways and forwards. The sideways forces cancel out and the overall movement is therefore forwards (see Fig. 10.8a).

(a) Propulsive action of tail (caudal fin)

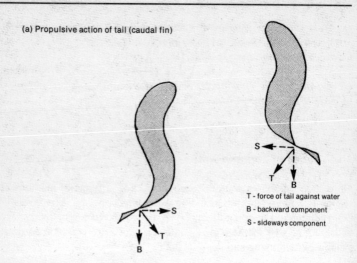

T - force of tail against water
B - backward component
S - sideways component

(b) Antagonistic action of muscle blocks

Anterior (head) end

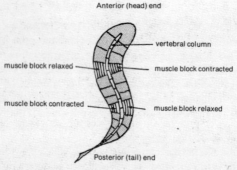

vertebral column

muscle block relaxed

muscle block contracted

muscle block contracted

muscle block relaxed

Posterior (tail) end

Fig. 10.8. *Propulsion in fish*

Stability: the fins give the fish a very precise steering capability and provide stability. The vertical fins, the dorsal, ventral and tail (caudal) fin prevent the fish **rolling** and **yawing** (see Fig. 10.9). The paired pectoral and pelvic fins prevent **pitching** and act as brakes to slow down or stop the fish (Fig. 10.9).

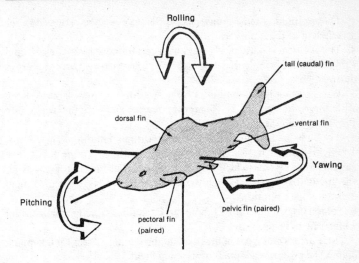

Fig. 10.9. *Types of displacement experienced by a fish*

Buoyancy: most bony fish are almost weightless in water and so, unlike birds, do not have to expend much energy in overcoming the force of gravity. They can use most of their energy in moving through the water. This weightlessness (buoyancy) is achieved by the possession of an air-filled space within the body, the **swim bladder**. The amount of gas in the bladder can be regulated. If the swim bladder is filled with air, the fish becomes less dense and will float higher up in the water. If the amount of air in the bladder is reduced, the fish will become denser and will sink lower down.

Bird flight, e.g. pigeon

Birds (page 340) have a number of major adaptations for flight:
1. The forelimbs are developed into wings and are covered in feathers, which are extremely light and provide a large surface area.
2. The pectoral muscles are developed into large flight muscles for raising and lowering the wings.

3. The sternum (breast bone) has a large keel (flat extension) for attachment of flight muscles.

4. The skeleton is particularly rigid with certain bones, e.g. the coracoid (collar) bones, being fused together for strength to withstand stresses during flight.

5. Weight is kept to a minimum by possession of a very light skeleton with many hollow bones. There are fewer bones in the wings (see Fig. 10.11) and there are no teeth.

6. The body is streamlined, with contour feathers smoothing the outline.

7. Flight muscles have a very high energy demand during flight. The gas exchange system (which has air sacs as well as lungs), together with the circulatory system, is highly efficient at delivering oxygen to flight muscles (for respiration).

8. A high body temperature (about 40 °C) enables rapid metabolism and a high rate of respiration.

There are several types of flight, including gliding, soaring and flapping flight. The pigeon is adapted for flapping flight.

Birds are heavier than air and in order to fly must generate lift. This is achieved by having an aerofoil-shaped wing. When the bird moves forward through the air, a high air pressure zone is created under the wing; this generates lift (see Fig. 10.10).

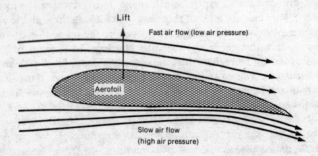

Fig. 10.10. *How an aerofoil generates lift*

In the pigeon's flapping flight, the downstroke of the wing is the power stroke and generates both lift and forward thrust. When the major pectoral muscles contract (Fig. 10.11) the wings are pulled downwards and slightly forwards, with the leading edge of the wing tilted down. On the upstroke (the recovery stroke) the minor pectoral muscles contract, lifting the wing and bringing it back to its original position.

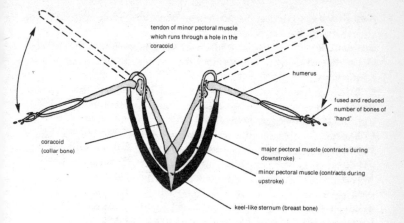

Fig. 10.11. *Front (anterior) view of pectoral girdle, wing skeleton and flight muscles*
NOTE. *The wing is an example of a pentadactyl limb but there is fusion
of bones and a reduction in the number of digits ('fingers').*

At the same time the wing is slightly folded and the feathers are tilted
to reduce air resistance which tends to push the bird downwards.

The wings and tail work together to steer and brake during manoeuvres
in flight.

10.6. *Support and Movement in Plants*

In herbaceous plants (page 320) support is achieved by a combination
of turgor (see page 60) and lignified supporting tissues. Parenchyma
(packing) tissue, when fully turgid, presses out on the inelastic epidermis.
Lack of turgidity, produced by drought, will cause a lack of support.
Under these conditions the plant will wilt (droop). Both xylem vessels
and sclerenchyma fibres, found in the vascular bundles, contain lignin,
and add mechanical strength to the stem, roots, leaves and leaf stalks.

In the stem of woody perennials (see page 320) support is entirely
achieved by lignified tissues. Each year a new layer of xylem is added
to the stem. This layer ceases to be functional in transporting water and
salts, and takes on a purely supportive role. The heartwood in older
trees is composed of xylem vessels and sclerenchyma fibres used solely
for support.

Plants, in general, show slow movements as a result of growth and
faster movements as a result of changes in turgidity. For details of growth
movements see pages 205–11.

Key Words

Locomotion
Hydrostatic skeleton
Exoskeleton
Cuticle
Bone
Calcium phosphate
Collagen
Ligaments
Tendons
Muscles
Axial skeleton
Skull
Cranium
Jaws
Vertebral column
Cervical vertebrae
Thoracic vertebrae
Lumbar vertebrae
Sacral vertebrae
Caudal vertebrae
Ribs
Sternum

Appendicular skeleton
Pectoral girdle
Scapula
Clavicle
Pelvic girdle
Pentadactyl limb
Immovable joints
 (sutures)
Slightly movable
 (cartilaginous) joints
Movable (synovial)
 joints
Ball-and-socket joints
Hinge joints
Sliding joints
Striped (skeletal)
 muscle
Smooth (visceral)
 muscle
Cardiac (heart) muscle
Antagonistic muscles
Flexor muscle

Extensor muscle
Biceps
Triceps
Muscle tone
Myotomes
Dorsal fin
Ventral fin
Caudal (tail) fin
Pectoral fins
Pelvic fins
Rolling
Yawing
Pitching
Swim bladder
Pectoral muscles
Aerofoil
Turgor
Xylem vessels
Sclerenchyma
Lignin
Heartwood

Exam Questions

1. (a) *List the names of the types of joints found in the skeleton and give*
one *example of where each type might be found.*   [5]
(b) *Make a large, labelled diagram to show the chief muscles and bones*
involved in movement at the human elbow.   [6]
(c) *Briefly explain why:*
(i) *the muscles shown in your diagram are often described as being 'anta-*
gonistic'
(ii) *each muscle has a nerve and blood supply.*   [5]

[CAM]

11. Reproduction

Definition: Reproduction *is the production of new individuals (offspring) derived from existing individuals (parents).*

Reproduction is a characteristic of all living organisms. There are two forms of reproduction, **asexual** and **sexual**.

11.1. *Asexual and Sexual Reproduction*

Asexual reproduction has the following features:
1. New individuals (offspring) are formed from *one* parent.
2. All offspring are genetically identical to each other and to the parent.
3. Asexual reproduction involves the type of cell division called **mitosis** (page 272).

Asexual reproduction has two main advantages:
1. Under favourable conditions, large numbers of offspring can be produced in a short time.
2. The offspring, being genetically identical to the parent, are normally able to survive under the same set of conditions.

Sexual reproduction has the following features:
1. A new individual is formed from *two* cells (**gametes**) which fuse to form a cell called a **zygote**. The two gametes usually come from two different individuals of the same species.
2. The offspring are genetically different to the parents and usually to each other. This generates variation within the species (see page 283).
3. The type of cell division called **meiosis** (page 272) gives rise to the gametes.

Sexual reproduction has one very important advantage: the offspring, being genetically different to the parents, are able to survive under different sets of conditions. This enables a species to withstand changes in an existing environment and to move into new environments.

Now try the following questions:

Q. Distinguish carefully between sexual and asexual reproduction.   [4]
[LON]

Q. Give one *advantage of asexual reproduction and* one *advantage of sexual reproduction.*   [2]
[OXF]

11.2. *Types of Asexual Reproduction*

Asexual reproduction is common throughout the plant kingdom but only occurs in simpler animals. Some of the more common methods are:

1. **Binary fission:** division of a unicellular organism into two identical individuals which then separate, e.g. in bacteria (page 307), *Chlamydomonas* (page 310) and *Amoeba* (page 326).

2. **Spore formation:** cells divide by mitosis to produce spores (single cells enclosed in a resistant coat). These disperse from the parent and under appropriate conditions each spore can develop into a new individual, e.g. in *Mucor* (page 314).

3. **Budding:** an outgrowth (bud) develops from the body wall. The bud grows to form a daughter organism which detaches to become independent, e.g. in *Hydra* (page 328).

4. **Vegetative propagation:** (see Table 11.1 and Fig. 11.1): this is a common method of asexual reproduction in flowering plants. It occurs when part of the parent plant becomes detached and develops into a new individual. This type of reproduction is often associated with storage organs. In herbaceous perennials (page 319) these storage organs may also act as organs of perennation.

Vegetative propagation has a number of advantages:

(a) The offspring, being genetically identical to the parent, are able to survive under the same set of conditions.

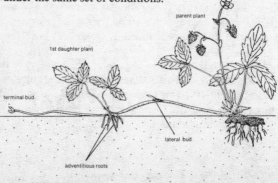

(a) A strawberry runner

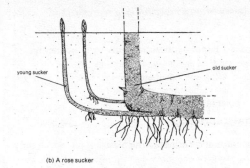

(b) A rose sucker

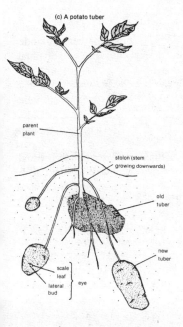

(c) A potato tuber

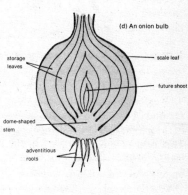

(d) An onion bulb

Fig. 11.1. *Vegetative reproduction*

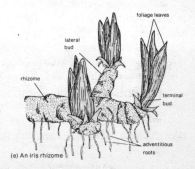

(e) An iris rhizome

Table 11.1. *Examples of organs of vegetative propagation*

Organ	Description	Examples
Runner	An overground, horizontal stem. If a bud comes in contact with damp soil, adventitious (false) roots develop and a daughter plant is formed. Eventually, the runner shrivels and the daughter plant becomes independent	Strawberry
Sucker	A short underground runner	Rose
The following are storage organs and organs of perennation:		
Stem tuber	An underground lateral stem which swells at the tip with food reserve. New plants arise from lateral buds on the tuber	Potato
Bulb	An underground stem growing vertically and surrounded by fleshy leaves containing food reserves. New plants arise from lateral buds at the base of fleshy leaves	Onion
Rhizome	An underground stem growing horizontally and swollen with food reserves. New plants arise from lateral buds on the rhizome	Iris

(b) The hazards involved in seed germination are avoided, e.g. the supply of food and water is provided by the storage organ or parent plant until the new plant is well established.

(c) Vegetative propagation tends to produce clumps of plants. This reduces the room for competition from other types of plants.

(d) As the offspring are genetically identical to the parents, specialized varieties and quality plants can be grown (propagated) without change. This is of great importance to the commercial grower.

5. **Artificial propagation:** man has developed artificial means of vegetative propagation by making buds form new plants. The two methods used are cuttings and grafts. **Cuttings** are short lengths of shoot which bear leaves and are planted in moist compost. They quickly develop adventitious roots and grow into new plants (e.g., *Geranium*). **Grafts** are used most extensively in the cultivation of rose and fruit trees. The variety of flower or fruit required is attached in the form of a twig, called the **scion**, to a well-established root system, called the **stock**. It is essential that the cambium of the scion and stock are in contact and that the joint between them is bound with tape to prevent infection and frost damage.

11.3. *Types of Sexual Reproduction*

Sexual reproduction occurs widely throughout plant and animal kingdoms, from unicellular organisms to mammals and flowering plants (see Table 11.2). Even bacteria carry out a form of sexual reproduction called conjugation.

Remember the main points about sexual reproduction:
1. It involves fusion of two gametes to form a zygote.
2. The production of gametes involves meiosis.
3. The offspring are genetically different to the parents.

The body cells of most organisms contain two sets of chromosomes, one set derived from the male parent and one from the female parent. This is called the **diploid** number of chromosomes (2n).

Definition: The diploid *number (2n) is the number of chromosomes in a body cell. It consists of two sets of similar chromosomes.*

We know that in sexual reproduction a new individual is formed by the fusion of two cells called gametes. If these cells had the normal chromosome complement of body cells, the zygote which results would have twice the diploid number of chromosomes (4n). This is not the case. The gametes contain only *half* the chromosome number of body cells, so that when they fuse they restore the normal chromosome number. The gametes have the **haploid** number of chromosomes (n).

Definition: The haploid *number (n) is the number of chromosomes in a gamete. It consists of one set of chromosomes.*

When male and female gametes fuse at fertilization, the diploid chromosome number is restored:

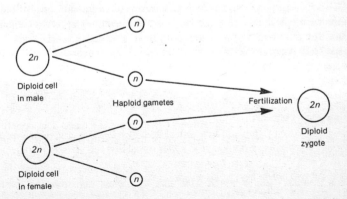

Definition: Fertilization *is the fusion of male and female gamete nuclei to form a zygote.*

In simpler organisms such as *Spirogyra* (page 312) and *Mucor* (page 314) the gametes are not produced by two distinct sexes, but rather there are a number of strains. Sexual reproduction occurs between different strains.

In more complex organisms two different sexes, male and female, are recognized. Certain parts of the body, the **reproductive organs** or **gonads**, have become specialized for the production of gametes. The **egg** is the gamete produced by the female. It is larger than the male gamete, is immobile and contains a food reserve. The **sperm**, the gamete produced by the male, tends to be much smaller and moves to meet the female gamete. In animals and lower plants the sperm is motile (moves by itself), whereas in flowering plants the male gamete is carried by external agents such as wind or insects.

In some organisms called **hermaphrodites**, individuals have both male and female reproductive organs and produce both kinds of gamete (see Table 11.2).

Definition: A **hermaphrodite** *is an individual which has both male and female reproductive organs and is able to produce both male and female gametes.*

Being hermaphrodite has two major advantages:
1. An individual can mate with any other adult individual of the same species and does not have to wait for an individual of the opposite sex.
2. In the absence of a partner, a hermaphrodite can potentially self-fertilize. However, in most hermaphrodite organisms, self-fertilization is prevented since this rather defeats the main object of sexual reproduction – to generate variation.

Tapeworms (page 330) and certain flowering plants do self-fertilize, but most hermaphrodites have mechanisms for preventing self-fertilization (see *Hydra*, page 328; the earthworm, page 333; and flowering plants, page 319). Answer the following question using the text and Table 11.2 to help you:

Q. (*a*) *From each of* four *different major groups of organism name* one *organism which is hermaphrodite (i.e. has both male and female sex organs in the same individual).*   [4]
(*b*) *From each of* four *different major groups name one organism which normally reproduces both sexually and asexually and describe in detail both processes of reproduction in* one *of these.*   [4, 13]
(*c*) *Give* two *advantages of hermaphroditism.*   [2]
(*d*) *Give* one *advantage of asexual reproduction and* one *advantage of sexual reproduction.*   [2]

                                                                                          [OXF]

Table 11.2. *The occurrence of asexual and sexual reproduction in plant and animal kingdoms*

Plants		Asexual	Sexual	Two Sexes	Hermaphrodite
Chlamydomonas	(page 310)	Yes	Yes	Strains	No
Spirogyra	(page 312)	Yes	Yes	Strains	No
Mucor, and Phytophthora	(page 314) (page 315)	Yes	Yes	Strains	No
Fern	(page 317)	Yes	Yes	Sometimes	Usually
Herbaceous plant	(page 324)	Yes	Yes	Sometimes	Usually
Deciduous tree	(page 325)	Yes	Yes	Sometimes	Usually
Animals					
Amoeba	(page 326)	Yes	No	—	—
Hydra	(page 328)	Yes	Yes	No	Yes
Tapeworm	(page 330)	No	Yes	No	Yes
Earthworm	(page 333)	No	Yes	No	Yes
Butterfly and Bee	(page 335) (page 337)	No	Yes	Yes	No
Fish	(page 337)	No	Yes	Yes	No
Frog	(page 339)	No	Yes	Yes	No
Bird	(page 340)	No	Yes	Yes	No
Mammal	(page 342)	No	Yes	Yes	No

In a number of groups, sexual reproduction is associated with the production of a resistant stage which can survive under adverse conditions. Examples are:

1. the zygospore of *Spirogyra* (see page 313)
2. the zygospore of *Mucor* (see page 315)
3. the seeds of annual flowering plants (see page 244)
4. the pupae of insects (see page 336).

In sexually reproducing organims fertilization may be **external**, occurring outside the body of the female, or **internal**, occurring within the female's body.

External fertilization is common in aquatic (water-living) organisms where water provides the medium in which sperm can swim to eggs. It also prevents gametes, zygotes and developing offspring from drying out.

Internal fertilization is a feature of terrestrial (land-living) organisms where there is little free water and little opportunity for male gametes to swim freely.

In both higher plants and vertebrates there is a trend towards internal fertilization in terrestrial (land) environments. Ferns (page 317) and mosses live on land, have external fertilization, and are dependent on water for transference of the male gamete to the female reproductive organs (page 318). Flowering plants have internal fertilization and are *not* dependent on water for transferring of the male gamete (see page 243).

Among the vertebrates (page 305), fish (which are obviously aquatic) have external fertilization, and so do amphibia (they return to water to breed). Birds, reptiles and mammals have internal fertilization.

In fish and amphibia large numbers of eggs are laid and fertilization rates are relatively low. The fertilized eggs are usually given little or no protection, there is little parental care, and the young hatch out at a relatively undeveloped stage. Mortality rates are high, hence the large number of eggs produced.

In terrestrial vertebrates, internal fertilization replaces external fertilization, a suitable aquatic medium being provided within the body. There is also the problem of offspring drying out and different strategies have been developed to prevent this happening. In birds, the embryo is enclosed in a chalky shell together with a large food store. The eggs are usually incubated (kept at a constant temperature) by the parents and the young birds after hatching are fed until they can fly and fend for themselves. In mammals, the developing offspring are retained within the body of the female and the offspring are born in a relatively advanced stage of development. After birth the offspring receive intensive parental care.

Internal fertilization is associated with a high degree of protection for developing offspring, together with parental care after birth or hatching. The greater the energy expended in protecting and caring for the young, the fewer the number of offspring produced, and the lower the mortality rate. This pattern is evident in Table 11.3.

Table 11.3. *Comparison of reproduction in fish, birds and mammals*

Fish, e.g. herring (see page 337)	*Bird*, e.g. pigeon (see page 340)	*Mammal*, e.g. rabbit (see page 342) and human (see page 249)
1. Mating usually seasonal	Mating usually seasonal	Mating usually seasonal (not in humans)
2. External fertilization	Internal fertilization (in the oviduct)	Internal fertilization (in the oviduct)
3. Thousands of eggs produced at one time	Few eggs produced at one time	Few eggs produced at one time (in humans, one)
4. Small amount of yolk in the egg	Egg contains large amount of yolk	Little or no yolk with egg (embryo supplied with food via placenta)

Fish	Bird	Mammal
5. Development of embryo usually external and unprotected	Development of embryo external but within egg shell, incubated and protected by parents	Development of embryo protected within uterus of mother
6. Young hatch from egg, but usually unprotected	Young hatch from egg, but protected	Mother gives birth to live young
7. No parental care (usually)	Parental care. Young are protected, fed and taught	Parental care. Young are protected, suckled and taught
8. Chance of completing development low	Chance of completing development high	Chance of completing development high

11.4. *Sexual Reproduction in Flowering Plants*

In a flowering plant, the **flower** is the organ of sexual reproduction. Angiosperms (page 319) do not exhibit the alternation of generations shown by mosses and ferns (page 317). In reality, however, the gametophyte and sporophyte generations are present, but in a disguised form. The pollen and ova represent the male and female gametophytes, while the flower and rest of the plant represent the sporophyte generation.

Mosses and ferns are dependent on water for the transport of the male gamete to the female gamete and for successful fertilization. Flowering plants have overcome this need. In effect, fertilization takes place internally, within the ovule of the flower. The male gamete is contained within the pollen grain which is resistant to drying out. It reaches a female sex organ on the wind or is carried by an insect.

Spores are the dispersal phase in mosses and ferns whereas in flowering plants the seed and fruit are.

Sexual reproduction in flowering plants takes the following pattern:

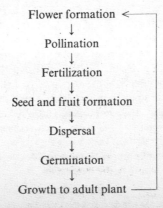

Flower formation ⟵
↓
Pollination
↓
Fertilization
↓
Seed and fruit formation
↓
Dispersal
↓
Germination
↓
Growth to adult plant ⟶

Flower structure

Flowers are very varied in shape and structure but all conform to a general plan. This is best shown by looking at a simple flower such as the buttercup. You are required to know the structure, and the functions of the parts, of *one* insect-pollinated flower (see Fig. 11.2).

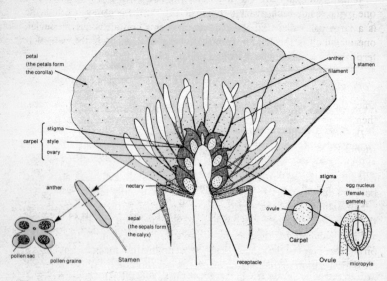

Fig. 11.2. *Structure of a buttercup flower* (*insect-pollinated*)

The parts of the flower are attached to the **receptacle**, the tip of the flower stalk. The parts are arranged in four rings, or **whorls**.

The outermost whorl is called the **calyx**. It is made up of five **sepals** – leaf-like structures which protect the developing flower bud. They fold back when the flower blooms.

The next whorl is called the **corolla**. It is made up of five **petals**. In insect-pollinated flowers such as the buttercup the petals are brightly coloured to attract insects. Some also have scented petals with a **nectary** at the base which produces sugary nectar. This also attracts insects.

Inside the corolla is the **androecium**. This is the male part of the flower and consists of a number of **stamens**. Each stamen is made up of a **filament** supporting an **anther**. Inside the anthers are pollen sacs in which the **pollen grains** (containing **male gametes**) develop.

The innermost whorl is the **gynoecium**. It is the female part of the flower and is made up of the female reproductive organs, the **carpels**. Some plants, such as the buttercup, have separate carpels, whereas others have a number of carpels fused together, as in the primrose. Each carpel consists of a wide, hollow base called the **ovary**, above which is a narrow region, the **style**. This ends in a sticky platform, the **stigma**, which receives pollen grains from the same or another plant during pollination.

Within the ovary are varying numbers of **ovules**. The buttercup has one ovule inside each ovary (see Fig. 11.2). At the centre of each ovule is a large cell called the **embryo sac**, and this contains several nuclei, one of which is the female gamete, or **egg nucleus**. It is this which is fertilized by a male nucleus from a pollen grain.

Pollination

Once the reproductive organs, the stamens and/or carpels, are mature, then **pollination**, the first stage of sexual reproduction, can proceed.

Definition: Pollination *is the transfer of pollen grains from anthers to stigmas.*

Pollination between anthers and stigmas of plants of the same species usually leads to fertilization. When 'accidental' pollination occurs between different species, fertilization does not usually result.

Self-pollination is the transfer of pollen from anthers to stigmas of the same flower, or between flowers on the same plant.

Cross-pollination is the transfer of pollen from one plant to the stigmas on another plant of the same species. Most pollination mechanisms ensure cross-pollination so that mixing of genetic information from two different plants occurs (remember, the main advantage of sexual reproduction is to produce variation in the offspring, see page 231).

Most flowering plants are hermaphrodite (have both male and female

Table 11.4. *Comparison of features of insect- and wind-pollinated flowers*

Insect-pollinated	Wind-pollinated
1. Large, brightly coloured petals	Small petals, not brightly coloured
2. May have nectar	No nectar
3. May be scented	Not scented
4. Small anthers (inside flower)	Large anthers (outside flower) which release pollen into the wind
5. Large spiny, sticky pollen grains produced in small number (to adhere to bodies of insects)	Small, light and smooth pollen grains produced in large quantities (only a very few will reach a stigma)
6. Sticky stigma inside the flower to obtain pollen from insect	Large feathery stigma(s) hanging outside the flower to trap pollen on the wind

sex organs) and, indeed, most plants have hermaphrodite flowers. Self-fertilization is usually prevented by the male flower parts, the stamens, maturing before or after the female parts, the carpels. When the stamens mature before the carpels, e.g. in the buttercup, it is called **protandry**. Much less common is **protogyny**, where the carpels mature first, e.g. in the bluebell. In some species there are separate male and female flowers, e.g. in the willow tree, and in others separate male and female plants, e.g. in the cucumber plant.

In most flowers cross-pollination involves one of two agents, insects or wind, to transfer pollen from one flower to another. Which particular agent is used can be determined by the appearance of the flower. The differences between wind- and insect-pollinated flowers are a common feature in exams (see Table 11.4). An oat (Fig. 11.3) is an example of a wind-pollinated plant.

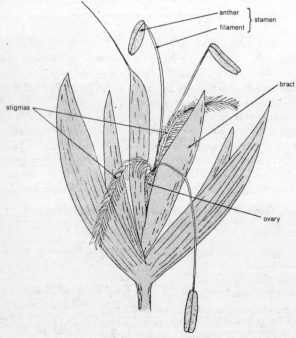

Fig. 11.3. *An oat flower (wind-pollinated)*

Fertilization

Once a pollen grain has arrived at the stigma of a flower of the same species, the male gamete has yet to meet the female gamete. Fertilization only occurs when the two gametes fuse.

The path between the male and female gamete nuclei is cleared by a remarkable process. Sugary secretions on the stigma stimulate the pollen grain to produce a pollen tube. The tube grows down towards the ovule (probably guided by chemicals) and then enters the ovule by a tiny hole, the **micropyle**. The tip of the pollen tube opens forming a clear pathway through which the male gamete reaches the female gamete (see Fig. 11.4).

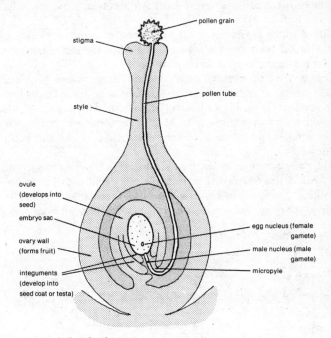

Fig. 11.4. *Just before fertilization*

The male gamete is a nucleus from one of the cells inside the pollen grain. This nucleus passes down the pollen tube into the embryo sac and brings about fertilization by fusing with the egg nucleus.

Fruit and seed formation

After fertilization the ovary undergoes changes to produce the **fruit**, the structure which contains and protects the **seeds**. The ovule develops into the seed and the ovary wall becomes the fruit. The other parts of the flower, the stamens, petals and sepals, are no longer needed: they shrivel up and drop off.

Inside the seed, the zygote divides many times by mitosis to become an embryo plant. This consists of:

1. a young root (the **radicle**)
2. a young shoot (the **plumule**)
3. one or two seed leaves (**cotyledons**). Monocotyledonous plants have one seed leaf, dicotyledonous plants have two (see page 319).

Stored within the seed are food reserves, either in the seed leaves, e.g. in the broad bean, or in a mass of cells called the endosperm which surrounds the embryo, e.g. in the maize-seed.

Outside the embryo plant, membranes called **integuments** harden and lignify to form a tough protective seed coat, or **testa**, around the embryo.

In summary, a seed is made up of an embryo plant and stored food, enclosed within a protective testa. Most of the water is withdrawn from the seed making it hard and extremely resistant to cold and other adverse conditions.

The main functions of the fruit are to protect and aid dispersal of the seeds. A **true fruit** is derived from the ovary wall and its form varies according to species. In beans and peas the fruit is the pod (see Fig. 11.5); in the plum, the skin, flesh and stone form the fruit; in an orange,

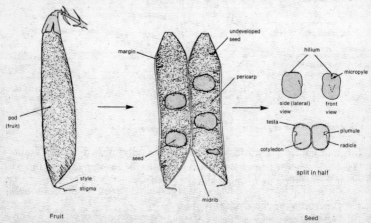

Fig. 11.5. *Fruit and seed of a pea*

the rind and flesh are the fruit, and in a hazelnut the hard shell is the fruit. **False fruits** occur where the ovary wall remains undeveloped. Instead, the receptacle enlarges to form the fleshy part attached to or enclosing the seeds, e.g. in apples, pears and strawberries.

The form the fruit takes is usually linked to the method of dispersal.

Fruit and seed dispersal

As plants are generally fixed, they must employ effective methods to disperse their offspring. This avoids overcrowding and competition for space, light and water, and also enables the spread of plants into new and different environments. The three main types of dispersal are:

1. **Wind dispersal:** the fruits of certain trees such as the ash or sycamore (see Fig. 11.6) develop wing-like structures. The fruits of certain herbaceous plants such as dandelion (Fig. 11.6) or willow herb develop into parachutes. In both types of fruit, the development of a large surface area relative to volume slows the fall of the fruit to the ground and produces dispersal over a wider area.

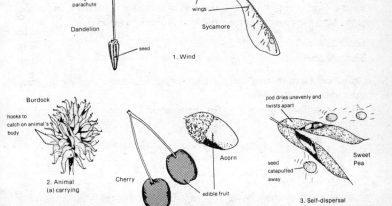

Fig. 11.6. *Methods of seed dispersal*

2. **Animal dispersal:** there are two main types:
(a) The fruit or seed may develop hooks which catch on the fur of animals, e.g. the fruit of the burdock (Fig. 11.6).
(b) The fruit may become fleshy (succulent), brightly coloured and sweet tasting. Small seeds, e.g. those of the blackberry or strawberry, can be

ingested and pass through the animal's digestive system relatively un-harmed. In such cases, the seed may be carried some distance before being deposited on the ground along with a convenient supply of fertilizer in the faeces. In other fruits, e.g. the cherry, the seeds in their hard coats are discarded after the soft fruit has been eaten.

3. **Self-dispersal:** the fruits of several plants such as the sweet pea, gorse, lupin and geranium develop into a pod or similar structure which, when it dries, catapults seeds away from the parent plant (see Fig. 11.6).

Accidental dispersal: all kinds of seeds can be dispersed accidentally in mud picked up on the feet of birds or mammals, or in man-made transport systems, e.g. trains, boats and aeroplanes.

Seed dormancy

Except in the case of ephemerals (page 319), dispersed seeds rarely germinate (grow into a seedling) immediately on reaching a suitable site. The period of inactivity prior to germination is called **dormancy**. Most seeds are produced towards the end of the summer and then have to undergo a period of dormancy (to allow for winter) before they will germinate. It is not always clear what triggers germination but in many cases a **period of time at a low temperature** followed by an increase in temperature is essential. The temperature change probably 'indicates' to the seed the onset of spring and favourable conditions.

Germination

Definition: Germination *is the growth of the seed into a seedling (a young plant able to photosynthesize).*

The conditions necessary for germination, following a period of dormancy, are:

1. water supply
2. suitable temperature
3. oxygen supply.

In addition, certain strains of grass seed require light before they will germinate. Most seeds, however, are indifferent to light intensity.

Before growth can occur insoluble food reserves, notably proteins and starch, must be broken down to soluble foods. Water is a reactant in these hydrolysis (breakdown) reactions and the enzymes that catalyse them require water and a suitable temperature. Oxygen is required for aerobic respiration to supply the energy for anabolic reactions.

The requirement for these conditions for germination to occur can be shown in a simple experiment using cress seeds (see Fig. 11.7). In

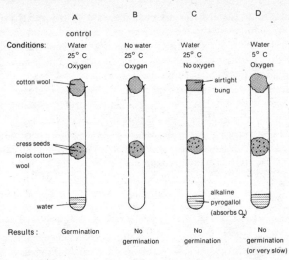

Fig. 11.7. *Experiment to investigate the conditions necessary for germination to occur*

this experiment, germination occurs only in the control seeds – those seeds provided with water, oxygen and a suitable temperature. If any of these factors are missing (as in tubes B, C and D), the seeds do not germinate. The conclusion is that oxygen, water and a suitable temperature are required for germination to occur.

During germination water is absorbed through the micropyle and the seed swells and the testa splits. Enzymes begin to convert insoluble food stores into soluble breakdown products. These are transported to growing points where they are utilized. Glucose (derived from starch) is used in respiration to provide energy for growth or is condensed to form cellulose used in new cell walls. Amino acids (derived from proteins) are used to manufacture enzymes and the structural parts of cells.

Q. What three *external conditions are essential for seed germination? Give one reason why each condition is essential.* [6]

[LON]

Germination in the broad bean

The main stages of germination can be observed in a large seed such as that of the broad bean (Fig. 11.8).

The bean seed is dicotyledonous (has two seed leaves) and the seed leaves contain a large store of starch and protein.

After the seed swells and the testa splits, the radicle (young root) grows

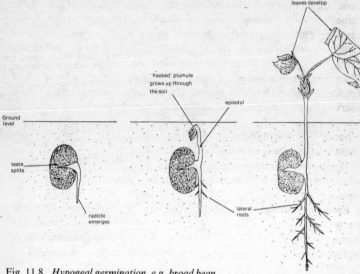

Fig. 11.8. *Hypogeal germination, e.g. broad bean*

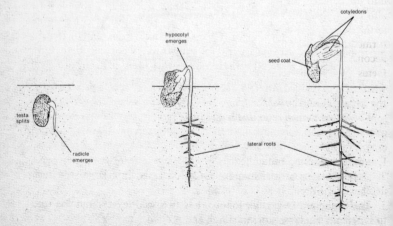

Fig. 11.9. *Epigeal germination, e.g. french bean*

from the region of the hypocotyl down into the soil by positive geotropism (page 205). The plumule (young shoot) grows from the epicotyl region upwards by negative geotropism. The young root develops root hairs (page 323) which absorb water and mineral salts from the soil. Later lateral roots grow off the radicle and provide anchorage for the seedling. When the plumule breaks through the ground surface young leaves develop and photosynthesis begins. The cotyledons shrink as their food reserve is used up and the leaves take over food production.

Types of germination
There are two types (see Figs 11.8 and 11.9):
1. **Hypogeal germination**, e.g. in the broad bean. The cotyledons remain below ground. Rapid elongation of the epicotyl forces the plumule above ground.
2. **Epigeal germination**, e.g. in the french bean. The cotyledons rise above the ground as the hypocotyl elongates. The cotyledons protect the plumule as it pushes up through the soil and they turn green and act as the first leaves.

11.5. *Mammalian Reproduction*

In mammals, the reproductive organs are the paired **testes** in the male and the paired **ovaries** in the female. The egg (ovum) is fertilized internally. During **copulation** (the act of mating) the erect **penis** of the male is inserted into the female's **vagina**. Sperm from the male are squirted (**ejaculated**) into the vagina and swim up the female's **oviducts** where a sperm may fertilize an egg (**ovum**) if one is present. After the gametes fuse, the zygote formed undergoes mitotic cell division (page 272) and differentiates to become an embryo. Once organs are visible in the embryo it is called a **foetus**. The foetus remains in the female reproductive structures and when fully developed (after 40 weeks in humans), the foetus passes out of the female during birth.

We will use human reproduction as our example: this follows the same overall plan as that in other mammals, but differs in the following points (compare with the rabbit, page 342):
1. The uterus of the human female is single, not branched.
2. Only one egg is normally shed at ovulation. Only one embryo develops at a time.
3. Development of the embryo takes a long time (40 weeks) in relation to the size of humans as mammals.
4. The reproductive cycle in women is called the menstrual cycle and

Table 11.5. *Comparison of sexual reproduction in mammals and flowering plants*

Differences:

Mammals	Flowering plants
1. Sexes separate	1. Usually hermaphrodite
2. Reproductive organs present throughout life	2. Reproductive organs temporary (usually renewed annually)
3. Male gamete – active, transferred by mating, swims to female gamete	3. Male gamete – passive, transferred by insects, or wind. Male nucleus reaches egg nucleus via pollen tube
4. Continual development of embryo	4. Embryo often enters a dormant period after initial development
5. Offspring motile (able to move about); no need for agents of dispersal	5. Offspring (seeds) dispersed by wind, animals or self-dispersal
6. Number of offspring low	6. Number of offspring (seeds) high
7. Survival rate among offspring high	7. Survival rate among offspring low

Similarities:
1. In both, sexual reproduction involves the fusion of *unlike* gametes.
2. In both, fertilization is *internal* (in the ovule of the flower, and in the oviduct of the female mammal).
3. In each case, the embryo begins development within the female reproductive structures.

lasts about 28 days. In most other mammals, if fertilization does not take place, the lining of the uterus is gradually reabsorbed. In the human female, the uterine lining breaks down and leaves the body as the **menstrual flow** (page 256).

5. Unlike other mammals, the human female does not have a period of 'heat' (highly receptive to males) at or around ovulation. Rather, the human female is potentially receptive at most times.

Reproductive organs

The male reproductive organs, the **testes**, lie outside the abdominal cavity in a sac called the **scrotum** (see Fig. 11.10). Here they are at a slightly lower temperature than the rest of the body – this favours sperm production. Within the testis, sperms are produced in coiled tubes called **seminiferous tubules**. Each tubule is lined with actively dividing cells which give rise to sperms (see Fig. 11.11). The tubules lead into the **epididymis**, where sperm are stored. From here, the **sperm duct (vas deferens)** leads to the **urethra** running through the **penis**. The urethra at different times carries urine (see page 172) and sperm.

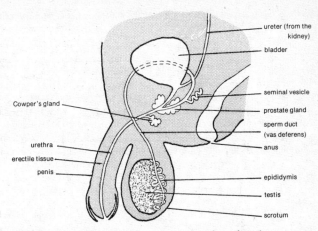

Fig. 11.10. *Human male reproductive organs (side or lateral view)*

Emptying into the vas deferens is the seminal vesicle which produces seminal fluid, the fluid in which sperms are suspended. The prostate gland produces secretions which activate the sperms to swim and provides them with nutrients. The penis contains many blood spaces which, during copulation (mating), become filled with blood at high pressure. This erects the penis, allowing it to penetrate the female's vagina during copulation.

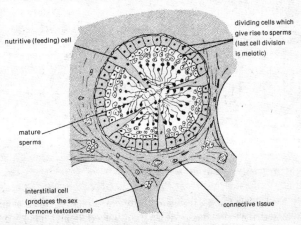

Fig. 11.11. *T.S. of a seminiferous tubule (highly magnified)*

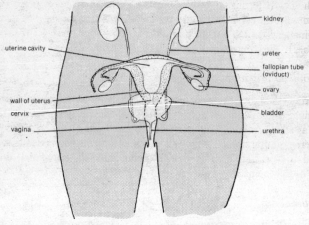

Fig. 11.12. *Human female reproductive organs* (*ventral view*)

The female reproductive organs, the **ovaries**, are two oval structures at the back of the abdomen (Fig. 11.12). They produce eggs (**ova**). From puberty onwards (see page 258) one or other of the ovaries releases an egg (ovum) about every 28 days. Inside the ovary, each ovum develops within a structure called the **Graafian follicle** (see Fig. 11.13). As the ovum matures the follicle moves towards the outer wall of the ovary.

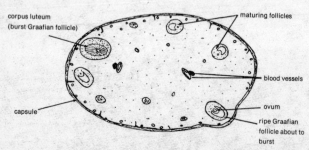

Fig. 11.13. *L.S. of ovary* (*slightly magnified*)

Finally, the follicle bursts, releasing the ovum, a process called **ovulation**. Once released, the ovum passes down the **oviduct (fallopian tube)** propelled by the beating action of cilia in the tube lining. Subsequent events depend on whether the egg is fertilized or not.

The oviducts lead into the wide muscular tube, the **uterus (womb)** where, if the egg is fertilized, development of the embryo takes place. A ring of muscle, the **cervix**, separates the uterus from the **vagina**. The muscular vagina leads to the outside and, during copulation, receives the sperms. At birth, the baby passes out through the vagina.

Copulation (*mating*)

During copulation the erect penis of the male is inserted into the female's vagina. Stimulation of sensory cells near the tip of the penis triggers off a reflex action causing muscular contractions of the epididymis, vas deferens and urethra. This results in **semen** (sperms suspended in seminal fluid and prostate gland secretion) being squirted (ejaculated) into the top of the vagina.

Fertilization

Active sperms swim through the cervix and up the oviducts, where they live for 2 or 3 days. A single ejaculation usually contains over 100 million sperm but few will enter the oviduct. If an ovum is present in one of the oviducts it may be fertilized by *one* of the sperms (see Fig. 11.14).

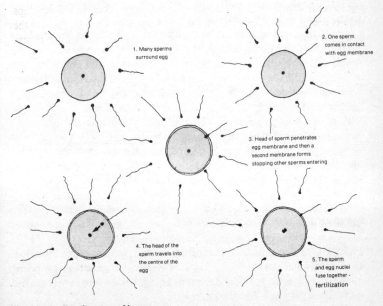

Fig. 11.14. *Fertilization of human ovum*

As an unfertilized egg lives for only about 24 hours, there are only 2 or 3 days in the month when fertilization is possible. Fertilization occurs when the head of the sperm penetrates the egg membrane and the nucleus of egg and sperm fuse. A membrane develops around the fertilized egg to prevent other sperm entering (see Fig. 11.14).

Gestation

The **gestation period** is the time from fertilization to birth and is around 40 weeks in humans. After fertilization the egg (now a **zygote**) divides many times by mitosis to become a hollow ball of cells. Now an **embryo**, it travels down to the uterus in about 4 to 7 days and then embeds itself in the uterine wall (**implantation**). Meanwhile, back in the ovary, the Graafian follicle which gave rise to the egg, has become an endocrine gland (see page 198) called the **corpus luteum**. This secretes the pregnancy hormone, **progesterone**, which stops further egg production by the ovaries and stimulates an increase in blood supply to the uterus.

As the embryo begins to form, finger-like projections called **villi** grow from it into the uterus wall. These, together with part of the uterine wall, become the **placenta**, a special organ which supplies the embryo with food and oxygen. By 2 months, the embryo has developed limbs and main organ systems and it is now referred to as a **foetus**. It is connected to the placenta by the **umbilical cord**.

In the placenta the foetal blood circulation comes in close contact with the mother's (maternal) blood circulation. Thin membranes separate foetal capillaries from blood spaces in the uterine wall and some substances can diffuse freely across (see Fig. 11.15). Oxygen and dissolved foods diffuse from the mother's blood to the foetal blood. Carbon dioxide and soluble excretory products such as urea, produced by the foetus, diffuse in the other direction. The villi give the uterus and placenta a greater surface area for the exchange of substances. The placenta also acts as a selective barrier preventing certain harmful substances reaching the foetus. In addition, the placenta produces the hormones **oestrogen** and **progesterone** which prevent menstruation (page 256) and maintain the blood supply in the uterine wall.

The foetus is surrounded by a 'water sac', the **amnion**, which contains **amniotic fluid**. This serves to protect the foetus from buffeting and sudden temperature changes, and allows the foetus to move.

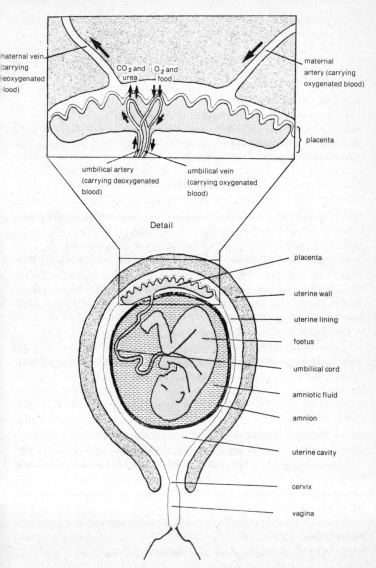

Fig. 11.15. *The foetus a few weeks before birth, with detail showing placenta*

Birth (*parturition*)

A few weeks before birth, the foetus comes to lie head down, towards the cervix. Just before birth, strong and rhythmic contractions of the uterine wall begin; this is the onset of labour. The cervix dilates (widens) to allow the baby's head to pass through and the amnion bursts releasing amniotic fluid. Muscular contractions of the uterus and abdomen eventually expel the baby head first through the vagina and into the world outside. The umbilical cord, still attached to the placenta, is cut and tied. A little later the placenta comes away from the uterine wall and is expelled as the 'afterbirth'.

Parental care

The baby is suckled on milk produced by the mother's **mammary glands**. These have been developing during pregnancy and milk production is stimulated by the baby suckling at the nipple. Milk contains nearly all the carbohydrates (lactose), fats, proteins, vitamins and mineral salts needed for the baby's growth and energy requirements. The milk lacks iron, but this has been stored in the baby's liver during pregnancy. At first the milk is the baby's only food, but this is gradually supplemented and later replaced by solid food.

In humans, the period during which offspring are dependent on parents is very prolonged compared to most other mammals. Parental care not only involves providing food, warmth and shelter, but also education, training and concern for psychological and moral welfare.

The menstrual (*oestrus*) cycle

NOTE. Details concerning how the cycle is controlled by hormones are not required by all examination boards. Check with your syllabus.

The reproductive cycle in women is known as the **menstrual** or **oestrus** cycle. Oestrus is the period of ovulation (when an egg is released). Menstruation is when the lining of the uterus is shed. The female reproductive cycle repeats itself about every 28 days from puberty (age 10–15) until menopause (at about age 50). The cycle is interrupted only by pregnancy.

The menstrual cycle, as shown in Fig. 11.16, is regulated by hormones secreted by the pituitary gland and ovary. During pregnancy, hormones secreted by the placenta delay recurrence of the cycle.

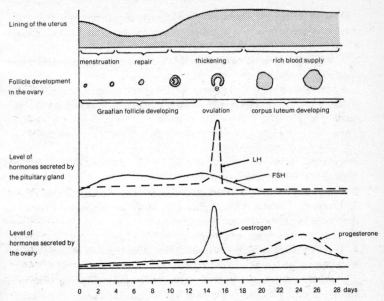

Fig. 11.16. *The menstrual cycle*

Normal cycle

Day 1: Menstruation begins and the pituitary gland (page 199) is stimulated to produce the hormone FSH (follicle stimulating hormone).

Days 5–14: FSH stimulates development of a Graafian follicle within the ovary. The follicle, in turn, secretes the hormone **oestrogen** which has a number of effects:

1. It causes the lining of the uterus to thicken (in preparation for implantation).

2. It inhibits the production of FSH by the pituitary gland (this, in turn, stops other follicles developing in the ovary).

3. It stimulates the pituitary gland to secrete the hormone LH (luteinizing hormone) – see below.

Day 14: Around this time the hormone LH stimulates the follicle to release its ripe egg (ovulation) which then travels down the oviduct.

Days 15–26: With the continued secretion of LH by the pituitary gland:

1. The burst Graafian follicle develops into a temporary endocrine gland, the **corpus luteum**.

2. The corpus luteum is stimulated to secrete the hormone **progesterone**, which builds up and maintains the lining of the uterus ready for implantation of a fertilized egg.

Days 26–28: In the absence of fertilization, the corpus luteum breaks down and the production of progesterone declines.This has two effects which start off the next cycle:

1. The lining of the uterus breaks down and blood and cell debris are discharged through the vagina (menstruation).

2. FSH production is no longer inhibited (there is no oestrogen or progesterone) so FSH is secreted and Graafian follicle development begins again.

Pregnancy cycle

If successful fertilization occurs:

1. The embryo implants in the uterus wall.

2. The corpus luteum persists, secreting progesterone, which maintains the uterus throughout most of pregnancy.

3. The mammary glands enlarge.

4. Later in pregnancy, the placenta takes over the role of the ovary in secreting progesterone and oestrogen.

Just before birth, the level of progesterone falls, allowing oestrogen and the pituitary hormone, **oxytocin**, to take effect and cause contractions of the uterus which expel the baby through the cervix and vagina.

After birth, a decrease in progesterone and oestrogen levels occurs (because the placenta is expelled) and this causes FSH to be released again by the pituitary gland. A return to the menstrual cycle follows.

Suckling causes another pituitary hormone, **prolactin**, to be released, and this causes secretion of milk from the mammary glands.

Notice how the reproductive cycle is governed by the pituitary gland and by feedback (see page 201) between four hormones, FSH, LH, oestrogen and progesterone.

Puberty

Puberty refers to the time of life when physical and psychological changes occur which transform a child into an adult capable of sexual reproduction. Puberty is triggered by hormones released from the pituitary gland. It is marked by the maturation of the reproductive organs and the development of secondary sexual characteristics. The reproductive organs (**gonads**) start to release the sex hormones, **oestrogen** (in females) and **testosterone** (in males). Secondary sexual characteristics refer to those physical features associated with sex (male or female) but not directly concerned with reproduction.

At puberty, the following physical changes occur:

In **girls** (age 10–15):
1. Ovaries mature and secrete oestrogen.
2. Menstruation starts.
3. Oestrogen stimulates development of the following secondary sexual characteristics:
(a) breast development
(b) fat deposits laid down under skin (giving the characteristic adult female body shape)
(c) hair grows under arms and in pubic region.

In **boys** (age 11–16):
1. Penis and testes enlarge, testes secrete testosterone.
2. 'Wet dreams' (nocturnal emissions of semen) begin.
3. Testosterone stimulates development of the following secondary sexual characteristics:
(a) body becomes more muscular
(b) larynx enlarges and voice deepens
(c) beard growth
(d) hair grows under arms and in pubic region.

Definitions

Reproduction
Haploid
Diploid
Fertilization
Hermaphrodite
Pollination
Germination

Key Words

Asexual reproduction	Internal fertilization	Style
Sexual reproduction	External fertilization	Stigma
Binary fission	Dormant	Ovule
Spore formation	Zygospore	Embryo sac
Budding	Gametophyte	Egg nucleus
Vegetative	Sporophyte	Self-pollination
propagation	Receptacle	Protandry
Runner	Calyx	Protogyny

Sucker
Stem tuber
Bulb
Rhizome
Artificial propagation
Cuttings
Grafts
Scion
Stock
Self-fertilization
False fruit
Wind dispersal
Animal dispersal
Self dispersal
Seed dormancy
Hypogeal
Epigeal
Testes
Ovaries
Copulation
Penis
Ejaculate
Egg (ovum)
Foetus
Menstrual flow
Seminiferous tubules
Epididymis

Sepals
Corolla
Petals
Nectary
Stamens
Filament
Anther
Pollen grains
Carpels
Ovary
Sperm duct
  (vas deferens)
Urethra
Graafian follicle
Ovulation
Oviduct (fallopian
  tube)
Uterus (womb)
Cervix
Vagina
Semen
Gestation
Embryo
Implantation
Corpus luteum
Progesterone
Villi

Insect-pollinated
Wind-pollinated
Micropyle
Fruit
Seed
Radicle
Plumule
Cotyledon
Testa
True fruit
Placenta
Umbilical cord
Oestrogen
Anmion
Anmiotic fluid
Mammary glands
Menstrual (oestrus)
  cycle
Oestrus
Menstruation
FSH
LH
Oxytocin
Prolactin
Puberty
Testosterone

Exam Questions

1. How does asexual reproduction occur in
(i)   Amoeba
(ii) a named mould fungus
(iii) a potato plant   [3, 4, 5]
                                              [O&C]

  (See pages 233, 314 and 326.)

2. Make a labelled diagram to show the structure of the flower of a named plant.   [9]
                                              [OXF]

3. (*a*) *Describe sexual reproduction in a* named *insect-pollinated flower referring in your answer to*
(*i*)   *pollination*
(*ii*) *fertilization, and*
(*iii*) *seed and fruit formation.*   [16]
(*b*) *In what ways does a wind-pollinated flower differ from your named example?*   [5]

[LON]

4. *Describe how the mammalian embryo is*
(*i*) *protected, and*
(*ii*) *provided with nutrients.*   [12]

[LON]

5. *Name four structures through which sperms may pass from the time that they are formed in the testes to the time that they are released from the body of a male mammal.*   [4]

[OXF]

12. Growth

A characteristic feature of living organisms is that they grow.

Definition: Growth *is an increase in the amount of protoplasm in an organism.*

The term protoplasm is used to distinguish between true growth and a simple increase in size due, perhaps, to an organism simply absorbing a lot of water.

Growth occurs when anabolic processes exceed catabolic processes (see page 35). Growth requires raw material and energy (from respiration).

The term 'growth' can be applied to an individual or to a population of individuals, in which case growth is increase in numbers.

In multicellular organisms, growth is accompanied by cell division, an increase in cell numbers, and an increase in the organism's complexity.

12.1. *Growth Curves*

Populations of organisms, and individual organisms themselves, show a pattern of growth based on, or a slight modification of, the standard growth curve (Fig. 12.1). This curve describes well the growth and decline

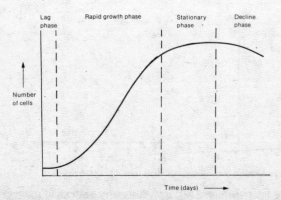

Fig. 12.1. *Standard growth curve for a microbial culture*

in numbers normally found in a culture of micro-organisms such as bacteria, protozoa, yeast or single-celled algae. During the **rapid growth phase** the population is increasing at a maximum rate for the species under the existing conditions. As they grow, the cells use up nutrients and produce waste products which accumulate in the medium. Eventually growth slows down and stops due either to a lack of nutrients or the build up of waste products (or both). This is the **stationary phase**. It leads into the **decline phase** in which the number of living cells in the population gradually decreases.

The standard growth curve is variously modified in multicellular organisms. The growth curve of an annual flowering plant (page 319) follows closely the standard form (see Fig. 12.2). In a perennial tree,

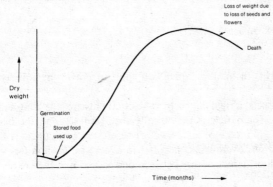

Fig. 12.2. *Growth curve for an annual flowering plant*

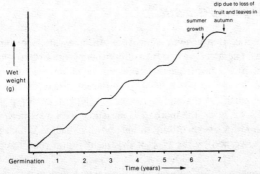

Fig. 12.3. *Growth curve for a perennial tree, e.g. oak*

e.g. the oak tree (page 319), growth stops in winter and resumes in spring (see Fig. 12.3). Loss of fruit and leaves in autumn accounts for a loss

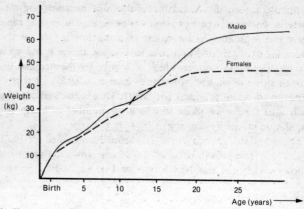

Fig. 12.4. *Human growth curves*

in weight. In humans, there is a spurt of growth at puberty (see Fig. 12.4).

In insects, a rigid cuticle (exoskeleton) prevents any continual increase in size. The cuticle is periodically shed (**ecdysis**) and the insect swells in size (by intake of air or water) before the new cuticle is laid down. As a result, overall size increases in a stepwise fashion (see Fig. 12.5). Growth of tissues occurs between one ecdysis and the next, so that the insect 'grows to fit' its new cuticle (Fig. 12.5).

12.2. *Measuring Growth*

You will notice that growth is measured using a variety of units of size or number. The ideal way to measure growth would be to measure the increase in protoplasm over a given period of time. However, this is difficult to do. Instead, a number of different measurements are used: increase in fresh (wet) weight, dry weight, length, or for populations, numbers. In all cases, the increase is measured over a set period of time. Each method has its own limitations and drawbacks. Which technique is used is governed by the type of organism(s) and the time period involved.

1. **Fresh weight:** this is an easy technique which does not involve injury to the organism. The whole organism is weighed at intervals. In plants, soil should be excluded. The fresh weight method does not always give an accurate indication of growth since fresh weight is influenced by

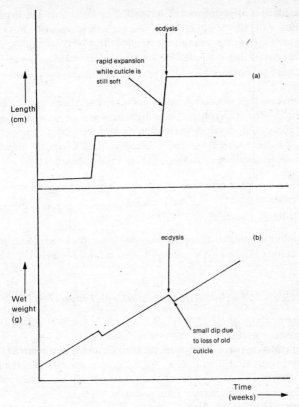

Fig. 12.5. *Growth curves for an insect showing incomplete metamorphosis, e.g. locust*

changes in water content of the body. It is used to measure growth in vertebrates.

2. **Dry weight:** this method usually gives a more meaningful result. It involves killing the sample by gently heating it to a constant weight to remove water. It is frequently used for seedlings. However, an increase in dry weight could be due to an increase in non-living material, e.g. lignin in plants. Also, it is possible for organisms to lose dry weight while still growing, as in the case of germinating seedlings which use up their food stores (see Fig. 12.2).

Valid results can be obtained using the dry weight method by taking large *random* samples at set time intervals from a large population kept under identical conditions.

3. **Length:** this straightforward method does not injure the organism. It is most suitable for organisms whose growth is channelled in mainly one direction, e.g. the young root or shoot tips of flowering plants. Its disadvantage is that it ignores growth in other directions, such as width. This may be considerable, for example, the increase in girth of tree trunks.

4. **Numbers:** this method is used for unicellular organisms such as bacteria, protozoa and algae. These organisms may be grown in solutions called **cultures**, from which samples are taken. The most common technique is to take several samples of fixed volume, count the number in each sample, and then extrapolate for the total population:

$$\frac{\text{average number of organisms per sample}}{\text{volume of sample}} \times \frac{\text{total volume}}{\text{of culture}} = \text{total population number}$$

Sampling is repeated at regular intervals to calculate the rate of growth of the culture. The results can be expressed graphically as in Fig. 12.1.

12.3. *Growth Patterns in Mammals and Flowering Plants*

Growth patterns in mammals

1. Growth takes place throughout the body, but may be **allometric** (some parts growing faster than others). In the human, for example, the brain grows slowly and stops growing at about age five. The legs, however, grow faster and keep growing until maturity.

2. Growth involves cell division and an increase in cell number, but not usually an increase in cell size.

3. Growth is **limited**, i.e. growth stops after a certain length of time. Growth processes may still occur, e.g. the production of new blood cells, but there is no overall increase in size.

4. Growth is, within limits, independent of temperature.

Growth patterns in flowering plants

1. Growth is usually restricted to regions called **meristems**. **Apical meristems** occur at root and shoot tips (see Fig. 9.16, page 209). The **cambium** is a circle of actively dividing cells in stems and roots which is responsible for their increase in girth (see Fig. 15.11, page 320).

2. Cell division is followed immediately by rapid cell elongation due to the rapid uptake of water (Fig. 9.17, page 209). This is followed by **cell differentiation** (the cells becoming specialized for particular functions).

. Growth is **unlimited**, i.e. a plant may retain the ability to increase its size throughout its life.

. Environmental factors, e.g. seasonal temperature variations, influence growth rate.

Now try the following question:

. *Briefly compare growth in a mammal with that in a flowering plant.*

[6]

[OXF]

Similarities in growth between mammals and flowering plants

. Growth is under the control of chemical substances (hormones).

. The rate and extent of growth depends, in some measure, on the balance and quantity of food. In plants this will be affected by the supply of minerals, water and light, and on the temperature.

. The growth curve for a mammal may resemble the same general form as that for a flowering plant (e.g. Fig. 12.1).

. Growth is determined genetically (the genes influence growth through the action of enzymes and hormones) – see also page 48.

12.4. *Metamorphosis*

Definition: Metamorphosis *is a major change in body form between the egg and adult stages of a life cycle.*

The intermediate forms between egg and adult are called **larvae**.

Metamorphosis is found in insects (page 335) and amphibians (page 339). Insects fall into two groups based on their type of metamorphosis. Insects such as locusts and cockroaches show **incomplete metamorphosis** (rather misleadingly named). This means the larval form gradually changes into the adult over a series of moults when the cuticle is shed

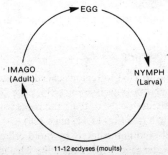

Fig. 12.6. *Incomplete metamorphosis in the cockroach*

(Fig. 12.6). At the final moult the reproductive organs are mature and the wings are fully formed. The larval stages are called **nymphs**. Insects such as butterflies, moths and true flies show **complete metamorphosis**. A larval form strikingly different from the adult hatches out from the egg. The larva feeds actively and grows rapidly, shedding its cuticle several times, before changing into a **pupa** or **chrysalis**. It is during this apparently inactive stage that metamorphosis occurs; there is an extensive breakdown of larval tissues and reorganization into adult tissues. Eventually the adult or **imago** emerges from the chrysalis (see Fig. 12.7 and page 335).

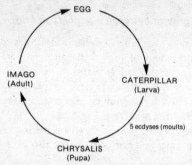

Fig. 12.7. *Complete metamorphosis in the butterfly*

In the butterfly, the larva is the major feeding phase of the life cycle, whereas the adult is the reproductive and dispersal phase. The two forms are specialized for different modes of life. The advantage of this is that there is no direct competition between larva and adult for space or food.

In amphibia, metamorphosis occurs when tadpoles develop into frogs or toads. Amphibians are only partially adapted for life on land and must return to water to breed. Metamorphosis is the transition between the aquatic larval phase, the tadpole, and the adult terrestrial phase, the frog or toad (see page 339).

Definitions

Growth
Metamorphosis

Key Words

Growth curve	Allometric	Incomplete
Rapid growth phase	Limited growth	metamorphosis
Stationary phase	Meristem	Nymph
Decline phase	Apical meristem	Complete
Ecdysis (moult)	Cambium	metamorphosis
Fresh (wet) weight	Cell differentiation	Pupa (chrysalis)
Dry weight	Unlimited growth	Imago

Exam Questions

1. Describe how you would measure growth in an individual larva of a named species of insect. What are the problems you would encounter, how would you express your results and what conclusion would you expect to draw from them? [7]

[OXF]

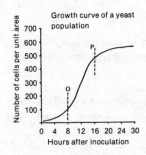

2. The curve X is typical of a population growth curve found in many organisms other than man.

(a) State one *fact about the yeast population up to point O on the graph that could explain the curve to that point.* [1]

(b) State one *fact about the yeast population between points O and P on the graph to explain the shape of the curve between these points.* [1]

(c) List three *factors which might cause the levelling-off as seen 15–24 hours after inoculation.* [3]

[SUJB]

3. *The graph shows the average body mass of boys and of girls up to the age of eighteen years.*

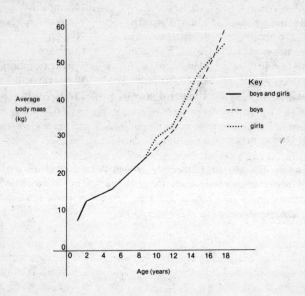

(a) *What is the average mass of six year old boys?*

(b) *Which have the greater average body mass (boys or girls)*

(i) *at age twelve?*

(ii) *at age eighteen?*

(c) *Suggest a reason for your answer to (b).*

(d) *What is the percentage increase in body mass of boys between age nine and age ten?*

[5]

[AEB]

13. Genetics

Genetics is the study of the mechanisms by which characteristics of parent organisms are passed on to their offspring. The material which carries the instructions for building a new organism is found in the **chromosomes** of the cell nucleus (see page 39). The substance within chromosomes called DNA (see page 48) determines which proteins are manufactured, and these in turn determine the characteristics of the organism.

13.1. *Chromosomes and Genes*

Chromosomes are coiled strands of protein and DNA present in the nucleus. They are visible with staining methods just before or during cell division. At other times they are dispersed in the nucleus as chromatin (see Fig. 3.3, page 38).

With the exception of reproductive cells (those giving rise to gametes), all cells in an organism's body have the same number of chromosomes (in man, 46; in the fruit fly, 8; and in rye grass, 20). In each of these cells there are actually two sets of chromosomes. This is called the **diploid** number of chromosomes (represented by 2n): in man 2n = 46.

Gametes contain half the diploid number of chromosomes: in man, sperm and eggs have 23 chromosomes. This is called the **haploid** number and is represented by n.

Chromosomes are normally found in pairs which have a similar appearance. These are called **homologous pairs**. For example in the body cells of man there are 23 homologous pairs. In sexually reproducing organisms one member of each pair is derived from one parent, and the other member from the other parent.

Chromosomes have two main roles:
1. They carry instructions for controlling day-to-day activities within the cell.
2. They carry instructions for producing a new cell (or in the case of a gamete, half the instructions for producing a new organism).

Each chromosome has a large number of **genes** running along its length. A gene is a short section of a chromosome (or more precisely, a short section of DNA). Each gene carries a coded message for manufacturing a specific protein. Genes control characteristics of the organism by providing instructions for manufacturing particular proteins. In a body cell, there are two copies of each gene (one on each chromosome of a homologous pair) so that a given characteristic, e.g. eye colour, is controlled by at least one pair of genes.

13.2. *Mitosis and Meiosis*

During cell division it is important that an accurate copy of the chromosomes, and the genes they carry, is passed on to daughter cells.

Mitosis is the type of cell division associated with growth and asexual reproduction. It occurs in somatic (non-reproductive) cells. A parent cell divides into two daughter cells which have identical chromosomes. The parental cell is diploid (2n) and the resulting daughter cells are diploid (2n). During the division process the number of chromosomes are doubled by **replication**. The total chromosome material then divides equally between the two daughter cells making them both identical (see Fig. 13.1).

Mitosis is a continuous process, but it can be broken down into a number of stages. For some examination boards you are required to know the different stages of mitosis. Check with your syllabus.

Meiosis is the type of cell division associated with sexual reproduction. It occurs in the cells which give rise to gametes. Meiosis has two important functions:

1. It produces daughter cells (gametes) which have half the chromosome number of the parent cell.

We have seen earlier (page 235) why this is so important for a sexually reproducing species. It is essential that gametes have half the chromosome number of a normal cell, so that when two gametes fuse to form a zygote, the normal (diploid) chromosome number is restored.

2. Meiosis produces daughter cells which have a combination of chromosomes different to that of the parent cell and to each other. This results in gametes which are genetically different and helps to produce genetic variation among offspring.

Meiosis generates variation by two processes:

(a) crossing-over

(b) the random assortment of chromosomes.

Meiosis involves two cell divisions, one after the other. This results in one parental cell giving rise to four daughter cells. Refer to Fig. 13.2 as you read through the text.

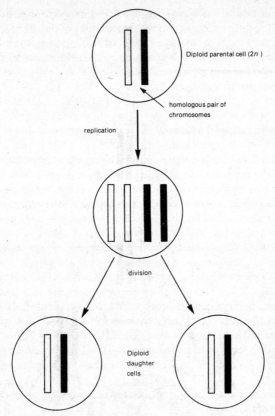

Fig. 13.1. *Mitosis (only one pair of homologous chromosomes shown)*

Before the first cell division, as in mitosis, the number of chromosomes is doubled by **replication** (duplication of chromosomes). In meiosis, however, homologous pairs of chromosomes come to lie side by side and attach at certain points called **chiasmata**. At these points genetic material can be exchanged between chromosomes by a process called **crossing-over**. The result is that a particular chromosome may come to contain some sections (and hence genes) from the male parent, and some from the female parent.

During the first division of meiosis, homologous pairs of chromosomes are separated out between two daughter cells. During the second or **reduction** division, these diploid cells divide to form haploid cells. These haploid cells form gametes.

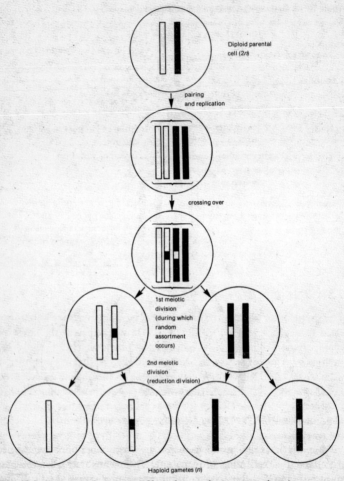

Fig. 13.2. *Meiosis (only one pair of homologous chromosomes shown)*

The second process which generates variation is the **random assortment of chromosomes**. During the first division of meiosis, which chromosome of a homologous pair goes to which daughter cell is determined at random. The result is that at the end of meiosis, each cell (gamete) contains a random assortment of chromosomes from each parent. For example, a human gamete has 23 chromosomes. All 23 chromosomes might come from one parent, or 10 from one and 13 from the other, or any other combination of the two which makes up 23.

Table 13.1. *Summary comparison of mitosis and meiosis*

	Mitosis	*Meiosis*
Number of cell divisions	1	2
Resulting cells	2 cells formed diploid (2n) identical	4 cells formed haploid (1n) non-identical
Purpose	Growth and replacement. In certain organisms (see page 237) asexual reproduction	Formation of gametes in sexual reproduction. Generation of variation
Occurrence	Growth regions (see page 266)	Gonads, e.g. testes and ovaries of mammals (see pages 250 and 252); anthers and ovules of flowering plants (see page 240). Sporangia of ferns (see page 318)

13.3. *Breeding Experiments*

The simple laws which govern inheritance of characteristics were formulated by Gregor Mendel in the 1850s. He worked with the sweet pea plant, *Pisum*. This is a hermaphrodite plant which normally self-pollinates. The characteristics Mendel worked with are controlled by what we now call **single-factor inheritance**. This means one gene controls one characteristic. Characteristics Mendel worked with included tallness of plant, colour of seed leaves (green or yellow) and texture of seed coats (smooth or wrinkled). In Mendel's time genes were unknown, but we can interpret the results of his experiments in the light of today's knowledge.

First of all Mendel established that his plants were pure-breeding for the characteristics he was investigating. This means that when the plant was 'selfed' (self-pollinated) the resulting offspring were identical to the parent with respect to the character being examined. For example, pure-breeding tall plants always produced tall offspring and pure-breeding dwarf plants always produced dwarf offspring.

To begin our investigation of inheritance let us examine one of Mendel's earlier experiments. When Mendel crossed a pure-breeding tall plant with a pure-breeding dwarf plant, and the resulting seeds were sown, all grew into tall plants. This first generation is called the F_1 (first filial) generation. When the F_1 plants were selfed or cross-pollinated, and the resulting seeds sown, the offspring, called the F_2 (second filial) generation, included tall

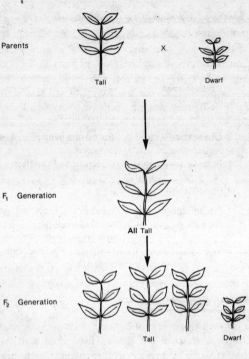

Fig. 13.3.   Ratio   3   :   1

plants and dwarf plants in the ratio 3 tall : 1 dwarf (see Fig. 13.3). Mendel called the tall character **dominant** and the dwarf character **recessive**. He carried out many experiments using these and other contrasting characters and repeatedly found that one character was dominant to another and gave the 3 : 1 ratio in the F_2 generation. We can interpret Mendel's experiment in terms of the transmission of genes from parent to offspring.

In an individual, a gene governing a particular characteristic will be present twice, once on each chromosome of a homologous pair. In an individual pure breeding for that characteristic, both copies of the gene will be the same. For example, in a tall plant we can represent the gene for tallness by T, and so the gene combination for the individual will be TT. In a pure-breeding dwarf plant, where the gene for dwarfness is represented by t, the gene combination is tt. Where two forms of a gene can exist, these forms are called **alleles**. T and t are thus alleles. It is usual to assign a capital letter to the dominant allele and a small letter to the recessive allele.

Definition: Alleles *are alternative forms of a gene, occupying the same site (locus) on a chromosome and affecting the same character.*

A gamete carries only one allele (because it is haploid and has only one of the chromosomes of the homologous pair). A TT individual will produce gametes carrying T, and a tt individual, gametes carrying t. If TT and tt individuals are crossed, a T gamete will fuse with a t gamete to produce a Tt offspring. Such F_1 offspring produce T and t gametes. If these offspring are crossed, and fertilization is random, then a variety of F_2 generation offspring will be produced carrying TT, Tt and tt allele combinations.

There is a standard procedure for writing out explanations for the results of genetics experiments. Let us apply this procedure to Mendel's experiment above.

Genotype refers to the genetic constitution of an individual. When applied to a particular gene, it indicates the combination of alleles. For example, a pure-breeding tall sweet pea plant has the genotype TT.

Phenotype is the observable characteristic of the organism resulting from its genotype, e.g. a TT sweet pea plant has a tall phenotype.

Use the first letter of the dominant character to name the alleles.

Let T = dominant allele for tallness

Let t = recessive allele for dwarfness.

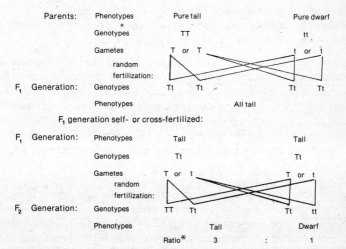

*NOTE. This ratio is only an expression of probability; it is the *expected* ratio of the different phenotypes. It does *not* mean that of four offspring produced three will be tall. This ratio is only likely, or a close approximation, when large numbers of offspring are considered.

When answering genetics problems set out your answers using the above format. For each genetic cross, write down the following terms as set out above:

Parents: Phenotypes

Genotypes

Gametes

F_1 Generation: Genotypes

Phenotypes

Then fill in the appropriate spaces with the data. Remember that gametes are **haploid** and carry only one set of genes whereas organisms are **diploid** and carry two sets of genes.

An individual which is pure breeding for a particular characteristic and therefore has two copies of the same allele is said to be **homozygous** for that character, e.g. TT. An individual which is not pure breeding for a character, for example with the genotype Tt, is said to be **heterozygous**.

When you have mastered the technical terms, and you can reproduce the format above, it is then a question of recognizing the type of inheritance. At O-level you need only be familiar with two types:
1. single-factor inheritance with complete dominance
2. single-factor inheritance with incomplete dominance.

In addition, on certain boards you are required to recognize sex-linked inheritance. You should ask your teacher about this.

13.4. *Single-factor Inheritance with Complete Dominance*

This type of inheritance can be recognized by:
1. the presence of only two phenotypes in the F_2 generation
2. a ratio of phenotypes in the F_2 generation of about 3 : 1.

The inheritance of tallness in sweet peas (see above) is an example of complete dominance. Other examples are:

Inheritance of brown and blue eye colour in humans (the allele for brown eye colour, B, is dominant, and the allele for blue eye colour, b is recessive).

Let us work through an example:

Q. When a red-flowered plant was crossed with a white-flowered plant all the offspring (F_1) had red flowers. When these offspring were crossed amongst themselves, 113 red-flowered and 38 white-flowered offspring (F_2)

were obtained. Using appropriate symbols, and with the aid of diagrams,
explain the genetics of these crosses as fully as you can. [14]
[LON]

This is clearly an example of single-factor inheritance with complete dominance:
1. There are only two phenotypes in the F_2 generation.
2. There is an approximate 3 : 1 ratio of phenotypes in the F_2 generation.

The allele for redness is dominant to the allele for whiteness.

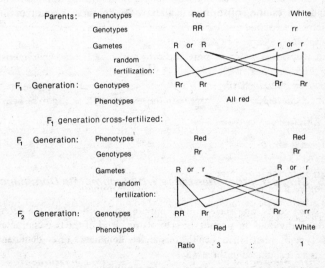

The test or back cross: clearly, under a system of complete dominance, a homozygous dominant individual (e.g. RR) and a heterozygous individual (e.g. Rr) have the same phenotype, i.e. they have the same physical appearance for that characteristic. However, their difference in genotype can be detected by the use of the back or test cross. The individual with a dominant phenotype is crossed with a individual with a recessive phenotype, a homozygous recessive (e.g. rr).

If this cross is carried out using a heterozygous individual with the dominant phenotype, the resulting offspring will yield dominant and recessive phenotypes in the ratio 1 : 1. For example, using tall and dwarf sweet pea plants:

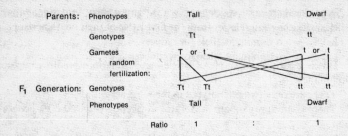

If the back cross is carried out using a homozygous dominant individual, the resulting offspring will all have the dominant phenotype:

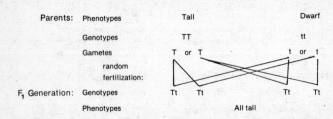

13.5. *Single-factor Inheritance with Incomplete Dominance*

For some genes, one allele is not dominant to another, but in a heterozygous individual the phenotype is a mix or blend of the two characters. This type of inheritance is called **incomplete dominance** or **co-dominance**. It can be recognized in three ways:

For a cross between two pure-breeding individuals:

1. The F_1 generation shows phenotypes which are intermediate to the parental phenotypes.
2. *Three* phenotypes are present in the F_2 generation.
3. The ratio of phenotypes in the F_2 generation is 1 : 2 : 1, where the intermediate phenotype is 2.

Let us look at an example: inheritance of coat colour in Shorthorn cattle.

When dealing with co-dominance it is conventional to designate both alleles with the capital letter from the first letter of their characteristic.

Let R = co-dominant allele for redness

Let W = co-dominant allele for whiteness.

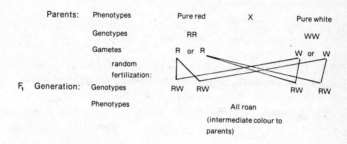

F₁ generation cross-fertilized:

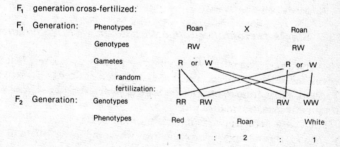

Let us try this out with a worked example.

Q. The seeds resulting from a cross between white-flowered and red-flowered parents produced a generation of pink-flowered plants. Explain, with the aid of appropriate symbols and diagrams, what you would expect the next generation to be if

(i)   the pink-flowered plants were allowed to self-pollinate

(ii) the pink-flowered plants were crossed with the white-flowered parent.

[17]

[LON]

This is clearly an example of incomplete dominance, since the off-spring show a mixture of parental phenotypes. The cross described in the question would be:

Let **R** = allele for redness
Let **W** = allele for whiteness.

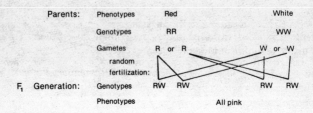

Now, in answer to part (i):
F_1 generation self-fertilized:

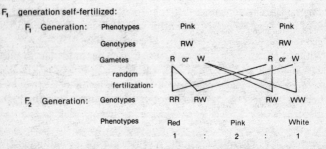

Offspring would be expected in the approximate ratio 1 red : 2 pink : 1 white, assuming large numbers were produced.

In answer to part (ii):

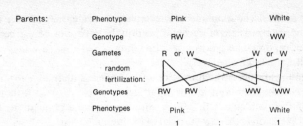

Offspring would be expected in the approximate ratio 1 pink : 1 white, assuming large numbers were produced.

Now turn to page 286 and try Questions 2–4. Remember, for each you must first decide whether inheritance involves complete dominance or incomplete dominance.

13.6. *Variation*

Every species has a set of characteristics which all members possess in common. The characteristics distinguish one species from another. This type of variation between different groups of organisms is the basis on which the classification system works (see page 302).

Within a species, clearly, variation exists from one individual to another and from one generation to the next. Variation is produced by a combination of the individual's genotype and the influence of environmental factors on the expression of that genotype. For example, basic body shape and size can be modified by environmental factors such as quantity of food eaten, exercise, disease, etc. Full expression of the genotype can be prevented or allowed depending on environmental conditions.

Sources of genetic variation

1. **Mutation:** this is a sudden change in either the amount or structure of chromosomal material (DNA). Mutations are of two kinds: **chromo-syndrome (mongolism) in man is a result of an individual having 47 chromosomes instead of 46. Gene mutations** involve a change in the syndrome (mongolism) in mammals is a result of an individual having 47 chromosomes instead of 46. **Gene mutations** involve a change in the structure of DNA. Haemophilia (blood failing to clot) is caused by a gene mutation.

Mutations occur spontaneously at a very low rate, but this rate may be increased by certain mutagenic agents in the environment, e.g. certain chemicals and various types of radiation (atomic radiation, u.v. light, and X-rays).

Mutations result in the formation of new alleles (alternative forms of a gene). Most mutations are harmful and recessive. However, they do provide the main source of genetic variation within a population. This genetic variation is the raw material of evolution.

2. **Genetic reassortment:** this results from the rearrangement of genes through

(a) crossing-over during meiosis (see page 273)

(b) the random assortment of chromosomes during meiosis (see page 274)

(c) the random fusion of gametes at fertilization.

Continuous and discontinuous variation

Genetically controlled characteristics generally fall into one of two categories: continuous or discontinuous variation.

1. **Continuous variation:** these are characteristics which show a continuous range of expression rather than clearly separated types. Height, weight and intelligence are examples of continuous variation. Such characteristics normally give a normal distribution curve. For example:

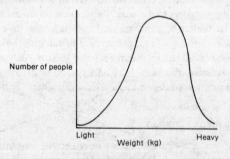

Number of people

Light   Heavy

Weight (kg)

Characteristics which show a continuous range of variation:

(a) are usually controlled by several genes (multiple-factor inheritance).

(b) can be influenced by environmental factors.

(c) have a low heritability (the characteristics are not clearly passed on to offspring).

2. **Discontinuous variation:** these are characteristics which show a discontinuous range of expression, i.e. types are clearly separated without a range of intermediates. Eye colour in humans and tallness in sweet pea plants are examples. Characteristics which show a discontinuous range of variation:

(a) are usually controlled by one or two pairs of genes.
(b) cannot be changed by environmental factors.
(c) have a high heritability (the characteristics are clearly passed on to offspring).

13.7. *Sex Determination in Humans*

Of the 23 pairs of homologous chromosomes present in human cells, one pair determines the sex of the individual; these are the sex chromosomes. In the female, both chromosomes have the same appearance and are called X chromosomes. In the male, one chromosome is the X chromosome while the other is shorter and is called the Y chromosome. Gametes, being haploid, will only carry one sex chromosome.

It is the sperm which determines the sex of the offspring. The arrangement ensures that on average half the offspring are male and half female:

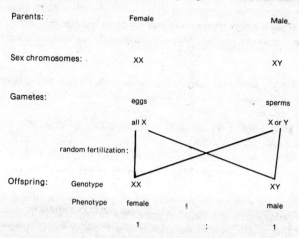

This method of sex determination is also found in other mammals.

Definition

Alleles

Key Words

Chromosomes
Genes
Diploid number
Haploid number
Homologous pairs
Mitosis
Meiosis
Replication
Crossing-over
Random assortment

Single-factor
  inheritance
F_1 generation
F_2 generation
Dominant
Recessive
Genotype
Phenotype
Homozygous
Heterozygous

Complete dominance
Incomplete
  dominance
  (co-dominance)
Chromosome
  mutations
Gene mutations
Continuous variation
Discontinuous
  variation

Exam Questions

1. (a) What are the differences between haploid and diploid cells? [4]
Give two *examples each of (i) haploid cells and (ii) diploid cells in mammals*
[4]
[LON]

2. (a) List the similarities and differences between meiosis and mitosis. [7]
(b) Explain, using appropriate genetic symbols, the results from genetic crosses between:
(i) a homozygous tall plant and a homozygous dwarf plant [6]
(ii) two heterozygous tall plants [6]
(iii) a heterozygous tall plant and a homozygous dwarf plant. [6]
In all these crosses assume that tall is dominant to dwarf.

[LON]

3. Industrial melanism derives from the presence of a gene B which leads to the development of a totally black colour in the adult of the peppered moth. This is dominant to the normal gene b which, in the homozygous state, results in a pale speckled form of the moth.
A black moth was mated with a pale speckled one and approximately equal numbers of black and pale speckled offspring were obtained.
(a) Work out the genetics of this cross. What can you deduce about the genotypes of the parents? [13]

(b) *What would have been the result of mating two heterozygous moths together? Show all stages in your working.* [12]

[OXF]

4. *In a genetic experiment a wild rabbit was mated with a tame white rabbit and it was found that all the offspring were brown, like the wild parent. These offspring were then mated with other white rabbits to produce a total of 216 offspring, 103 of which were white and 113 were brown. Explain fully the genetics of these crosses.* [13]

[OXF]

5. (a) *Briefly distinguish between the terms*
(i) *discontinuous variation, and*
(ii) *continuous variation.*
Give two examples of each. [6]

[O&C]

14. Ecology

Definition: Ecology *is the study of living organisms in relation to their surroundings.*

These surroundings are the **environment**, which is made up of living (**biotic**) and non-living (**abiotic**) components. The stable system formed by these components in a particular area is termed an **ecosystem**.

Definition: An ecosystem *is the living and non-living components of a region which interact to produce a stable system.*

Examples of ecosystems are a pond, a hedgerow and a beech wood. Ecosystems contain a number of **habitats** each with their own **community** of organisms.

Definition: A habitat *is a specific locality with a particular set of environmental conditions and its own* community *of organisms.* Put another way, a habitat is a place where an organism lives. In a pond ecosystem, the fringe (edge) of the pond is a habitat.

Definition: A community *is the collection of interacting species within a habitat.*

Within our fringe habitat, the common reed, *Phragmites*, the reedmace, *Typha*, the frog, *Rana*, and the mayfly, *Ephemera*, are members of the community.

14.1. *The Abiotic* (*Non-living*) *Environment*

The abiotic environment has both physical and chemical components. These conditions dictate which organisms may live in a given area and particularly affect the abundance of green plants, upon which animal populations depend. Some of the more important abiotic features are:

1. **Light:** solar radiation (sunlight) is the energy source for all life. It is required by green plants to make the organic substances on which all organisms depend (see page 63).

In flowering plants, day length coordinates events in the life cycle such as germination, flowering and fruit development. In mammals, day length triggers seasonal body changes like growth of winter coat, hibernation and reproductive behaviour.

2. **Temperature:** metabolic reactions, controlled as they are by enzymes,

only operate efficiently within a narrow temperature range, generally 4–40 °C. Organisms, in general, choose habitats which will allow them to maintain an optimum temperature.

3. **Water** is the 'universal solvent' – the medium in which all life processes occur. It is also a transport medium and is a substrate for a number of important metabolic reactions, e.g. photosynthesis. Organisms living in dry environments must have a means of conserving water, e.g. a thick cuticle in plants, or efficient kidneys in mammals.

4. **Inorganic materials:** the availability of certain inorganic materials will affect plant and animal growth, either as nutrients, e.g. calcium, or poisons, e.g. heavy metals, or by producing unfavourable conditions like extreme pH (high acidity or alkalinity).

14.2. *Chemical Cycles*

Whereas energy flows through an ecosystem and is eventually lost, chemical resources are recycled. Three important chemical cycles are those of carbon, nitrogen and water.

Carbon cycle

Carbon is found in nature as atmospheric carbon dioxide (0·04 per cent of air), in other inorganic molecules, e.g. carbonate, within organic molecules in living organisms, and within organic molecules outside living organisms, e.g. coal or man-made synthetic compounds.

Atmospheric carbon dioxide is the source of carbon for all green plants, and ultimately, for all animals. Photosynthesizing plants extract carbon and incorporate it into carbohydrates. Some of these carbohydrates are later converted into proteins and fats. Carbon is returned to the environment by respiration, decay or combustion (see Fig. 14.1).

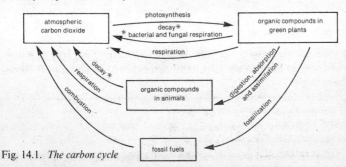

Fig. 14.1. *The carbon cycle*

Nitrogen cycle

Nitrogen is necessary for making proteins. About 80 per cent of atmospheric air is nitrogen. As a gas, it is inert (unreactive) and unavailable to most living organisms. However, nitrogen-fixing bacteria such as *Rhizobium* can use atmospheric nitrogen. They live symbiotically (see page 101) in the root swellings (nodules) of leguminous plants such as the pea or bean. The bacteria supply nitrogen (as nitrate) to the plant and receive sugar and protein in return. The plant uses the nitrogen for protein synthesis, animals obtaining their protein by eating plants. The nitrogen is recycled by decay of animal and plant material under the action of fungi and bacteria. **Nitrifying bacteria** convert ammonium compounds released by decay, first into nitrites and then into nitrates. Denitrifying bacteria reduce the nitrate content of soil (and hence reduce the soil's nutrient value) by converting nitrates into nitrogen gas. The nitrogen cycle is summarized in Fig. 14.2.

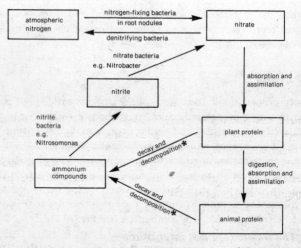

Fig. 14.2. *The nitrogen cycle*   ✳ by bacteria and fungi

Q. Soil bacteria play a vital role in the circulation of nitrogen in nature. Which one of the following bacterial processes is not *useful to flowering plants.*

A. *the oxidation of ammonium compounds to nitrites*

B. *the oxidation of nitrites to nitrates*

C. *the fixation of nitrogen from the air*

D. *the conversion of nitrates to nitrogen gas*

E. *the conversion of animal and plant protein to ammonium compounds.*

Water cycle

Water is circulated between the land, sea and atmospheric air (see Fig. 14.3). A small but significant portion passes through living organisms.

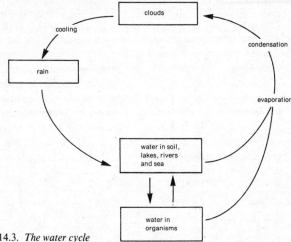

Fig. 14.3. *The water cycle*

Water is a raw material of photosynthesis (page 64) and is a source of hydrogen and oxygen for all living organisms. Plants and animals return water to the atmosphere in many ways, including excretion, transpiration (in plants), sweating (in mammals), and after death, by decay.

Questions on chemical cycles are popular with examiners (see for example, Question 1 on page 299).

14.3. *The Biotic (Living) Environment*

Within a community and its environment there is a continual flow of energy and matter. Energy enters the system as sunlight which is trapped by green plants and during photosynthesis is converted to chemical energy within organic substances (see page 64). One way or another, these substances are eventually broken down in respiration (page 105) and the energy released is lost from the system in the form of heat.

Green plants are **producers** of organic material whereas animals are **consumers**. In any habitat there are also **decomposers**, such as bacteria and fungi. They break down dead plant and animal material and return it as inorganic material back to the soil. In this way matter is recycled: producers convert inorganic material into organic matter and decomposers convert organic remains into inorganic material.

In any community **food chains** are built up, the producers (green plants) being eaten by herbivores (plant-eating animals) and the herbivores being eaten by carnivores (meat-eating animals). An example of a food chain is shown in Fig. 14.4. Each stage of the chain is known as a **trophic level**.

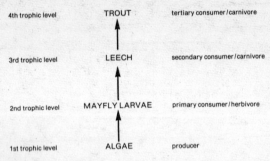

Fig. 14.4. *Food chain in a Welsh stream*

Food chains in this simple form rarely exist. Each consumer usually has several food sources and in turn is preyed on by several organisms. This results in **food webs** developing in any community. Fig. 14.5 shows a food web which incorporates the food chain in Fig. 14.4.

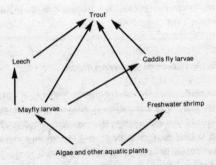

Fig. 14.5. *Part of a food web in a Welsh stream*

Only 1 per cent of the light energy that reaches a plant is incorporated into plant tissues. As energy is passed along the food chain, about 90 per cent is lost between one level and the next (see Fig. 14.6). For example,

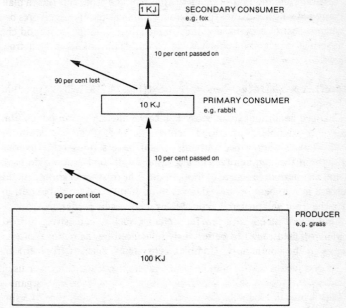

Fig. 14.6. *Energy flow through a food chain*
      (*values in boxes give energy content of tissue*)

of the energy required to maintain an individual rabbit, a large proportion is used in respiration to provide energy for movement, and to repair and maintain the body. This energy is eventually converted to heat. Only a fraction of the energy required to maintain a rabbit can be passed on to a rabbit's predator, the fox. The fox can only obtain energy from digestible material in the rabbit.

This decrease in available energy in a food chain explains why higher trophic levels have a lower **biomass** (mass of living material); there is less energy available to support living material. It is this energy loss at each level which makes animal protein so expensive to produce. It explains why Third World countries tend to rely more on crops than livestock. It also explains why the number of links in a food chain rarely exceed five.

In general, the higher the trophic level in the food chain, the larger the size of the individual organism (after all, a predator must overpower

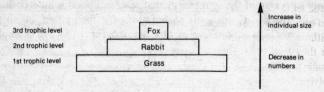

Fig. 14.7. *Pyramid of numbers*

its prey). **A pyramid of numbers** is produced (Fig. 14.7) as well as a pyramid of biomass.

Changed conditions in an ecosystem cause fluctuations in population sizes. For example, in our food web in Fig. 14.5, if the population of mayfly larvae were wiped out this would have serious repercussions throughout the community. The leech and caddis fly larvae would have to find an alternative source of food, or die. The freshwater shrimp might increase in numbers because there is more plant food available but, in turn, the trout will probably reduce their numbers by predation.

Simple ecosystems with few food chains, such as intensively farmed single-crop fields, tend to be seriously influenced by the removal of one species in the community. Complex ecosystems, such as tropical rain

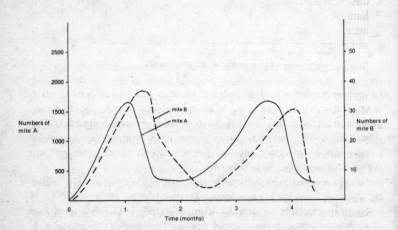

Fig. 14.8. *A predator–prey cycle for two species of mite*

forests, tend to be more stable, as consumers can find alternative prey when one type becomes unavailable.

Although populations show short-term changes in size, in the long term they tend to oscillate around average values, a situation called a **dynamic equilibrium**. Fig. 14.8 shows a predator–prey population cycle involving two species of mite (related to spiders). Mite B eats mite A. Mite A, which feeds on fruit, increases in numbers when food is plentiful. The population of mite B, which feeds on them, also increases. This results in a drop in the population of A, followed by a drop in the population of B, and the cycle then repeats itself. Notice, as you would expect, the numbers of mite A are higher than those of B.

14.4. *Soil*

Soil is a complex medium providing physical support and anchorage for plant roots and a supply of water and nutrients for plant growth.

Soil typically contains the following constituents:

1. mineral particles
2. water
3. air
4. dissolved mineral salts
5. humus
6. micro-organisms
7. soil animals.

1. **Mineral particles:** the character of soil is determined by the size and nature of the mineral particles it contains. This in turn dictates which plants and animals may live there.

Mineral particles are formed by the weathering of rocks. **Soil texture** refers to the size of individual particles in the soil. **Soil (crumb) structure** refers to how these particles are organized, i.e. how they clump together.

There are three main soil types:

(a) clay (particles less than 0·002 mm in diameter)
(b) silt (particles 0·002–0·02 mm in diameter)
(c) sand (particles 0·02–2 mm in diameter).

Sandy soils do not retain water well – the water evaporates or drains away. They have loose texture and give poor support to plant roots and are susceptible to wind erosion. Mineral salts are easily washed out (**leached**) from the soil.

Clay soils drain poorly, are heavy and are hard to work. The small particles are closely packed, hindering drainage and reducing aeration.

The soil (crumb) structure is governed by the soil texture (size of particles) and the proportion of dead organic matter (**humus**) present. The more organic matter present and the greater the mixture of different soil types, the more crumbly the soil.

A **loam** is a soil which contains a balanced mixture of sand and clay, and has a good humus content. It retains water but remains aerated and can be rich in mineral salts. Loams are usually excellent soils for growing plants.

2. **Soil water** exists as a thin film around soil particles and is held with an increasing force, the drier the soil becomes. The water content of a soil varies with soil type (see 1).

Available water is water that is available to be absorbed by plant roots. It includes capillary water (held to soils by capillary attraction) and drainage water.

Non-available water is bound more strongly to soil particles. Roots cannot exert a sufficient force to remove and absorb this water by osmosis (see page 152).

3. **Soil air:** air fills the spaces between soil particles except in a soil which is water-logged. Soils with large particles have more air spaces and drain rapidly. Soil air has more carbon dioxide and slightly less oxygen than atmospheric air because of aerobic respiration (page 107) by plant root tissues and other soil-dwelling organisms.

4. **Mineral salts** make up 0·02 per cent of the dry weight of soil. They originate mainly from the decomposition of organic matter (from plant and animal remains and excretory products). A range of these dissolved salts are essential for plant growth (see page 79). The balance of salts affects soil pH and therefore determines which plants can survive there.

5. **Humus** (dead organic matter) comprises the remains of dead organisms and their waste products. In soil used for growing crops humus rarely exceeds 6 per cent of the dry weight. Humus contributes to soil structure (see 1) and supplies nutrients and minerals as well as preventing their leaching out of the soil.

6. **Micro-organisms:** these include **saprophytic** bacteria and fungi (see pages 308 and 314). These break down organic compounds and humus to release soluble salts which are absorbed in solution by plant roots (page 159). **Nitrifying** bacteria convert ammonium compounds to nitrates (see page 290). **Nitrogen-fixing** bacteria (page 290) convert nitrogen gas to nitrate.

7. **Soil animals:** earthworms and beetles loosen the soil by burrowing, thereby increasing aeration and drainage. Earthworms contribute to soil fertility in a number of ways (see page 333).

NOTE. For certain examination boards you are required to know the experiments for determining soil composition, pH, and air, water or humus content. Check with your syllabus.

Q. (a) List the components of a fertile soil. [6]
(b) Give four ways in which earthworms improve soil fertility. [4]

[LON]

Soil fertility

Agricultural practices aim to increase soil fertility.

1. **Manuring:** continued harvesting of crops removes soil nutrients which must be replaced. Natural (organic) fertilizers like manure (animal faeces) contain most, if not all, essential nutrients for plant growth, particularly nitrogen. They decay slowly and provide humus which maintains or improves soil structure.

Artificial (inorganic) fertilizers are quick-acting and can be selected to provide particular nutrients, e.g. nitrogen, phosphorus and potassium (NPK). They do not, however, improve soil structure and are easily leached into rivers and lakes, where they cause unwanted plant growth.

2. **Liming:** lime adds calcium to the soil, neutralizes acid and causes clay particles to cling together (**flocculate**), improving aeration and drainage.

3. **Ploughing** in spring breaks up soil clumps to produce a good seed bed. It also aerates the soil increasing microbial activity.

In autumn, ploughing opens up the soil to aid frost penetration. This breaks up the soil and helps to kill pests and weeds. Ploughing is also used to mix manure with the soil.

4. **Crop rotation**, such as the Norfolk Four Course (see Fig. 14.9).

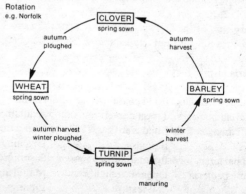

Fig. 14.9. *Cropping cycle*

Crop rotation achieves the following:

(a) It maintains the high nitrate content of soil. (NOTE. Clover is a leguminous plant, see page 290.)

(b) It reduces pest numbers (most pests are specific for a particular crop and so do not survive when the crop is changed).

(c) It helps keep the soil weeded.

(d) It is economical on manure.

(e) Different crops draw mineral ions from different levels of the soil.

Modern agriculture, through the use of artificial pesticides and fertilizers, need not rotate crops so rigorously, e.g. the intensive cereal cycle (see Fig. 14.10).

Fig. 14.10.
Intensive cereal cycle

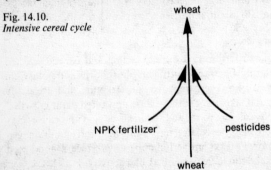

Definitions

Ecology
Ecosystem
Habitat
Community

Key Words

Carbon cycle	Food chain	Humus
Nitrogen cycle	Trophic level	Available water
Nitrogen-fixing bacteria	Food web	Non-available water
Nitrifying bacteria	Biomass	Leaching
Denitrifying bacteria	Pyramid of numbers	Saprophytic
Producers	Dynamic equilibrium	Manuring
Consumers	Soil texture	Liming
Decomposers	Soil (crumb) structure	Crop rotation
	Loam	

Exam Questions

1. Questions (a) to (e) refer to the numbered stages in the following diagram of part of the nitrogen cycle.

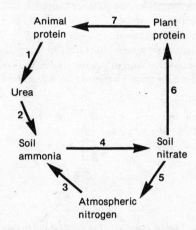

Which of the numbered stages represent the following processes?

(*a*) *Denitrification is*
A   2
B   3
C   4
D   5
E   7

(*b*) *Deamination is*
A   1
B   4
C   5
D   6
E   7

(*c*) *Nitrification is*
A   1
B   2
C   3
D   4
E   5

(*d*) *Nitrogen fixation is*
A   2
B   3
C   4
D   5
E   6

(*e*) *Decay is*
A   1
B   2
C   3
D   5
E   7

[5]
[LON]

2. *The diagram below shows some of the feeding relationships of inhabitants of a British oak wood.*

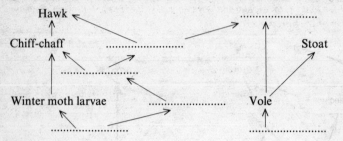

(a) *Insert the following organisms or parts of an organism in the appropriate spaces in the food web above.*

Oak tree leaves
Oak tree fruit (acorns)
Blue-tit
Aphid
Ladybird   [3]

(b) (i) *When a stoat eats a vole energy is transferred to the vole. Name three ways in which the stoat might use this energy.*   [3]
(ii) *Only a small proportion of the energy from the vole becomes incorporated into the stoat. What happens to the rest of the energy?*   [2]
(c) *Each autumn the oak trees shed their leaves. Explain how the nitrogen in the leaves is made available for the growth of trees in subsequent years.*   [7]

3. (a) *The diagram shows a simplified food web involving trout at different stages of its life history.*

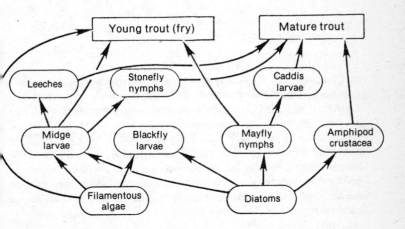

(i) *The ultimate source of energy for all the organisms in the food web is the sun. Describe* one *route by which the sun's energy may be made available for movement in trout. In your answer you should make reference to any processes important in the transfer of energy.* [8]

(ii) *Using examples from the above food web to illustrate your answer, explain what is meant by each of the following terms,*

A *producer*

B *primary consumer*

C *secondary consumer* [6]

(iii) *Describe* two *differences in the diets of young and mature trout.* [2]

(iv) *Suggest a possible effect on mature trout of a disease killing all mayfly nymphs. Explain your answer.* [3]

[AEB]

4. *List the living and non-living components of a well-balanced soil. Explain the contribution of each component to soil fertility.* [10, 15]

[LON]

15. The Diversity of Life

In Chapters 5 to 12 we examined the seven characteristics of life – nutrition, respiration, excretion, sensitivity, growth, movement and reproduction – largely as applied to mammals and flowering plants. These are the most complex representatives of the animal and plant kingdoms. However, on all syllabuses you are required to have some knowledge of a range of organisms. A selection of these organisms are included in this chapter. Representative species from the main groups of organisms are briefly described. Only those details which are emphasized on syllabuses are included. Check in the Contents list to find out which organisms are required by your examination board.

15.1. *The System of Classification*

The naming of living organisms follows a system of classification. This is really a way of organizing animals and plants into groups to help us describe and study them. There are three main advantages to the system of classification that is used:

1. It makes everyone talk the same language. Russian, Japanese and Australian biologists all understand that the scientific name *Canis familiaris* refers to the domestic dog.

2. It summarizes information. The name 'mammal' conjures up a picture of a hairy, highly developed animal, visible to the naked eye.

3. The system reflects evolutionary relationships between organisms. Several hundred years ago, organisms were grouped together according to seemingly related features. For example, earthworms, tapeworms and snakes were all classed as 'worms' because of their shape, and insects, birds and bats were placed in another group because they could fly. Nowadays, both animals and plants are grouped according to their evolutionary relationships. Snakes, for example, have a much more advanced body form than the tapeworm or earthworm. In terms of structure and physiology (function), snakes have much more in common with mammals, birds, fish and amphibians, and are included with these organisms in the group 'vertebrates' – animals with a vertebral column (backbone).

The groups that biologists use to classify organisms are known as **taxons**. The largest are **kingdoms**, of which there are two: the plant and animal kingdom. These are broken down into progressively smaller taxons, the members of which have more and more features in common. The smallest groups are called **species**.

Definition: A species *is a group of organisms with a large number of structural and physiological (functional) features in common. The members of a species can interbreed to produce fertile offspring.*

Horses are an example of a species. Horses and asses are different, but closely related, species. They can interbreed, but the offspring so produced, called mules or hinnies, are infertile.

All species of organisms are allocated scientific names according to the **binomial** ('two name') system introduced by Linnaeus in the eighteenth century. All species are given two names in Latin:

1. The first name is the **genus** name and is written with a capital letter, e.g. *Homo* (man).
2. The second name is the **species** name and is written with a small letter, e.g. *sapiens* (wise).

The full name is either written in italics or is underlined, e.g. *Homo sapiens*.

Any organism can be precisely filed away within the classification system. Table 15.1 gives two examples. A mnemonic is provided to help you remember the list of taxons in the right order.

At this level of study, you need only be able to classify organisms to the level of phylum, or in some cases (arthropods, vertebrates and seed-bearing plants), to the level of class. However, check with your syllabus.

Table 15.1. *Use of the classification system*

Mnemonic	Taxon	Examples	
		Man	*Bulbous buttercup*
Keep	Kingdom	Animal	Plant
Plucking	Phylum (pl. phyla)	Chordata	Spermatophyta
Chickens	Class	Mammalia	Angiospermae
Or	Order	Primates	Ranales
Face	Family	Hominidae	Ranunculaceae
Getting	Genus (pl. genera)	*Homo*	*Ranunculus*
Sacked	Species	*sapiens*	*bulbosus*

15.2. *The Plant Kingdom*

The term division is sometimes used in place of phylum.

Phylum Bacteria: microscopic unicellular organisms with simple structure (see page 307).

Phylum Algae: simple aquatic green plants. Many are unicellular. Examples are *Chlamydomonas* (page 310) and *Spirogyra* (page 312).

Phylum Fungi: non-green plants (no chlorophyll). Either saprophytic or parasitic. Body consists of threads (hyphae). Examples are *Mucor* (page 314) and *Phytophthora* (page 315).

Phylum Bryophyta: mosses and liverworts. Small green land plants with simple leaves and root-like structures. Show alternation of generations. The gametophyte generation is dominant to the sporophyte generation.

Phylum Pteridophyta: ferns – green plants which have roots, stems and leaves but do not bear seeds. Show alternation of generations. The sporophyte generation is dominant. Example, *Dryopteris* (page 317).

Phylum Spermatophyta: green plants with roots, stems and leaves and which bear seeds. Have a well-disguised alternation of generations. The sporophyte generation is dominant (see page 239).

1. **Class Gymnospermae** (gymnosperms) – cone-bearing plants.
2. **Class Angiospermae** (angiosperms) – flowering plants. Examples, the buttercup, *Ranunculus* (page 324) and the chestnut tree, *Aesculus* (page 325).

15.3. *The Animal Kingdom*

It can be divided into those animals without a vertebral column, 'invertebrates', and those with a vertebral column, 'vertebrates'. All phyla from Protozoa to Echinodermata are invertebrates. Most members of the phylum Chordata are vertebrates.

Phylum Protozoa: single-celled animals. Example, *Amoeba* (page 326).

Phylum Coelenterata: aquatic. Body made up of two layers of cells surrounding a cavity – the enteron – whose only opening is the mouth. Example, *Hydra* (page 328).

Phylum Platyhelminthes (flatworms): small flattened worms which lack a body cavity. Many are parasitic. Example, the tapeworm, *Taenia* (page 330).

Phylum Annelida (true worms of annelids): round worms with segmented bodies and through gut. Example, the earthworm, *Lumbricus* (page 333).

Phylum Arthropoda (arthropods): the largest animal phylum in terms of species. Possess hard exoskeleton, jointed limbs and jointed appendages.

1. **Class Crustacea** (crustaceans): body with many jointed appendages. Examples, shrimps, crabs and woodlice.

2. **Class Arachnida** (arachnids): body divided into two, with eight legs. Examples, spiders, mites and scorpions.

3. **Class Myriapoda** (myriapods): long body with many segments. Examples, centipedes and millipedes.

4. **Class Insecta** (insects): body divided into three, with six legs. Examples, bees, ants, flies, butterflies (page 335).

Phylum Mollusca (molluscs): soft-bodied animals often with a shell. Examples, slugs, snails, oysters, octopus.

Phylum Echinodermata (echinoderms): marine 'spiny-skinned' animals based on a five-rayed plan. Examples, starfish, sea urchins.

Phylum Chordata (chordates): all those below are vertebrates.

1. **Class Pisces** (fish): aquatic. Fins and gills present. Poikilothermic. Example, herring, *Clupea* (page 337).

2. **Class Amphibia** (amphibians): semi-terrestrial. Eggs laid in water. Moist skin. Poikilothermic. Example, frog, *Rana* (page 339).

3. **Class Reptilia** (reptiles): terrestrial. Eggs laid on land protected by leathery shell. Scaly skin. Poikilothermic. Examples, snakes, turtles.

4. **Class Aves** (birds): forelimb modified to form wing. Eggs protected by chalky shell. Skin covered with feathers. Homoiothermic. Example, pigeon, *Columba* (page 340).

5. **Class Mammalia** (mammals): give birth to young. Young suckled by mammary glands. Skin covered with hair. Homoiothermic. Example, rabbit, *Oryctolagus* (page 342).

15.4. *Viruses*

Viruses do not fit into either animal or plant kingdoms and are classified separately. Some biologists do not classify them as 'living' because they do not show *all* of the seven characteristics of life (see page 34). They can, however, reproduce.

Structure

They are **ultra-microscopic**, meaning that they cannot be seen using a light microscope, but only with the aid of an electron microscope. They are approximately 0·0001 mm in diameter which is about 10,000 times smaller than an *Amoeba*.

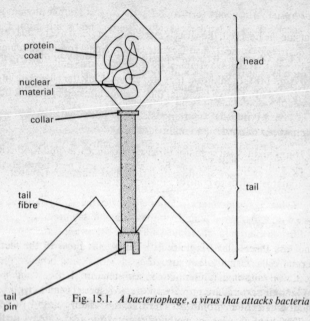

Fig. 15.1. *A bacteriophage, a virus that attacks bacteria*

They come in a variety of geometric shapes, including spheres, rods and hexagons. A bacteriophage virus is shown in Fig. 15.1.

They lack a nucleus, cytoplasm and cell membrane and are therefore **non-cellular**. The body consists of a **protein coat** surrounding a strand of **nucleic acid** (DNA or RNA).

Mode of life

Viruses are **parasitic**. In order to reproduce they must invade a host cell and take over its metabolic machinery to make copies of themselves. In so doing they burst and destroy the host cell. The damage viruses do to the host's tissues means they cause disease.

A particular virus normally has a specific host organism. Viruses are usually named after the disease they cause:

Name of virus	Host	Mode of transmission
Influenza (flu) virus	Man	Droplets (sneezing)
Poliomyelitis (polio) virus	Man	Water
Tobacco mosaic virus	Tobacco plant	Insects such as greenfly

Control

Outside the host cell, the virus is inert and some exist as crystals. They can be destroyed by heat, disinfectants or u.v. light. Inside the mammalian body, some viruses are attacked by antibodies in a similar way to bacteria (see page 148). In addition they are attacked by the chemical **interferon** (see page 148). Unlike bacteria, they are not affected by antibiotics.

Importance to man

1. **Harmful:** they cause major diseases in man, cattle and food crops.
2. **Helpful:** some viruses can be used to control pest populations, e.g. myxomatosis virus to control rabbits.

15.5. *Bacteria*

They are unicellular (single-celled) but do not have all the features of normal cells. Usually they are grouped with plants because they have a cell wall (although it does not contain cellulose).

Habitat: bacteria are probably more widespread than any other group and live in most environments on earth, the reason being:
1. They have a variety of methods of nutrition.
2. They can withstand a wide range of environmental conditions – different temperatures, pH, etc. Some bacteria form resistant spores.

Structure: bacteria have a diameter about 1/100 that of an *Amoeba*; they are just visible with a light microscope.

Unlike other cells they do *not* have membrane-bound organelles. For example, mitochondria are not present and the nuclear material floats 'free' in the cytoplasm.

Unlike plants they do *not* have:
1. a cell wall made of cellulose (instead it contains fat and protein)
2. a large vacuole.

Bacteria come in different shapes and sizes. Sometimes they group together in clumps or chains. They are classified according to shape:
1. spherical (coccus)
2. straight rod (bacillus)
3. curved rod (vibrio)
4. spiral (spirillum).

A generalized bacterium is shown in Fig. 15.2.

Nutrition: some bacteria are **autotrophic**, building up organic substances from simple inorganic substances, either using sunlight energy (photosynthesis) or the energy from the oxidation of inorganic sub-

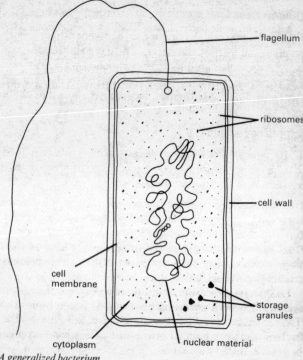

Fig. 15.2. *A generalized bacterium*

stances, e.g. nitrite to nitrate (chemosynthesis). Photosynthetic bacteria
do not contain chloroplasts; chlorophyll is scattered throughout the
cytoplasm.

Most bacteria are **heterotrophic** and require 'ready-made' organic food.
Of these, some are **saprophytic** and some are **parasitic**.

Saprophytic bacteria feed on dead organic material. Digestive enzymes
are secreted on to the substratum (surface on which the bacterium is
found), food is digested **extracellularly** (*outside* the cell), and the products
of digestion are absorbed across the cell wall.

Parasitic bacteria cause disease:

Name of disease	Host	Mode of transmission
Typhoid	Man	Eating or drinking contaminated food or water
Syphilis (venereal disease)	Man	Sexual contact
Tetanus	Man	Bacterial invasion of wounds
Pneumonia	Man	Water droplets in breath

Respiration: some aerobic, some anaerobic, some both.

Movement: some bacteria have flagella for locomotion (see Fig. 15.2).

Reproduction: asexual – by **binary fission** (dividing into two). Under favourable conditions a bacterium can divide once every 20 minutes, producing over one million offspring in less than 8 hours.

Some bacteria have a form of sexual reproduction called **conjugation**, whereby fragments of genetic material are exchanged.

Importance to man

1. **Useful:**

(a) Saprophytic bacteria recycle chemicals in nature (see Fig. 14.1 and 14.2):

(i)   **Nitrogen-fixers** convert atmospheric nitrogen into nitrates.

(ii) **Nitrifying bacteria** convert nitrogenous wastes into nitrates.

(iii) **Decay bacteria** break down dead organisms.

(b) **Cellulose-digesting** bacteria are used by herbivores which culture these bacteria in their gut (see page 101). This is an example of a symbiotic relationship.

NOTE. A **symbiosis** is an association between two organisms of different species where both partners derive benefit, frequently nutritional benefit. The rabbit benefits from the digestion of cellulose by the symbiotic bacteria; the same bacteria benefit from a continual supply of cellulose brought to them by the rabbit.

(c) **Vitamin-synthesizing** bacteria are found in the gut of man and other mammals.

(d) Certain bacteria are used in industrial food processes, e.g. making butter, cheese, yoghurt and vinegar.

(e)   **Genetic engineering:** by genetic manipulation certain bacteria can be induced to make otherwise scarce and expensive chemicals, e.g. insulin.

2. **Harmful:**

(a) **Disease bacteria**, living in plants and animals, damage host tissue and excrete by-products which are harmful to the host.

(b) **Denitrifying bacteria** (see page 290) remove nitorgen from the soil, thus reducing its fertility.

(c) **Saprophytic bacteria** cause decay (putrefaction) of food.

Control

Bacteria require moisture, a suitable temperature and a food supply in order to survive. They are readily killed when exposed to u.v. light (found in sunlight).

Foods can be preserved by providing an environment unsuitable for bacteria:
1. **Drying** foods, e.g. peas, raisins, milk, meat.
2. **Salting** foods, e.g. fish, ham (high osmotic potential plasmolyses bacteria).
3. **Pickling** foods, e.g. onions (pH too acid for bacteria).
4. **Freezing** foods, e.g. meat ($-20\,°C$ suspends life).

Live bacteria can be prevented from being ingested by:
1. **Boiling** food – this kills most bacteria, but not bacterial spores.
2. **Pressure-cooking** ('sterilizing' or 'autoclaving') food – this kills all bacteria including spores.
3. **Pasteurizing** (heating and rapidly cooling) milk.
4. **Chlorinating** drinking water.
5. **General hygiene** and cleanliness about the house and use of **disinfectants** (chemicals which kill bacteria).

If bacteria invade the body and cause infection they can be counteracted by:
1. White blood cells which engulf bacteria (see page 134).
2. White blood cells which produce antibodies (see page 134).
3. Antibiotics (see page 317) administered to the sufferer, e.g. penicillin.
4. Injections of antibodies (see page 149) administered to the sufferer.
5. If the infection is on the body surface, e.g. sore throat, antiseptics (mild disinfectants) can be used.

15.6. *A Simple Plant:* **Chlamydomonas**

Phylum (Division): Algae.

Habitat: freshwater ponds and ditches. In large numbers it colours the water green.

Structure: about 0·02 mm long. It shows a mixture of plant and animal features (see Fig. 15.3).

Plant features:
1. a cellulose cell wall
2. chloroplast (for photosynthesis)
3. pyrenoid (region of starch formation).

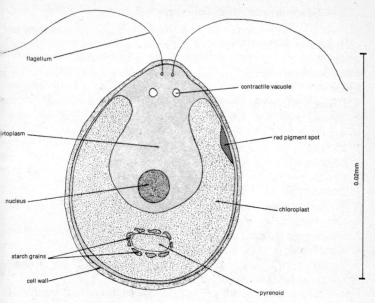

Fig. 15.3. Chlamydomonas

Animal features:
1. flagella (for locomotion)
2. pigment spot (to detect light and move towards it – phototaxis)
3. contractile vacuole for osmoregulation.
   Nutrition: autotrophic, by photosynthesis (see page 64).
   Respiration: aerobic (see page 107).
   Gas exchange: simple diffusion across cell wall. As with other green plants, in sunlight, carbon dioxide diffuses in and oxygen diffuses out. At night, oxygen diffuses in and carbon dioxide diffuses out (see page 79).
   Locomotion: using flagella.
   Sensitivity: moves towards light (phototaxis) and probably is sensitive to other stimuli, e.g. temperature, chemicals.
   Reproduction: asexual and sexual. Asexual, by fission – divides into four or more daughter cells. Sexual, when individuals of different mating types (strains) come together.

15.7. *A Filamentous Alga:* Spirogyra

Phylum (Division): Algae.

Habitat: floats in masses on or near the surface of still water such as ponds and canals.

Structure: *Spirogyra* is a filamentous alga (see Fig. 15.4) It consists of identical cells joined end to end, forming a long strand or filament. Individual filaments are visible to the eye as green, hair-like strands.

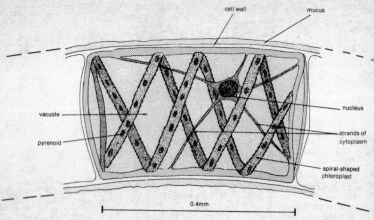

Fig. 15.4. *A* Spirogyra *cell*

Although *Spirogyra* is made up of many cells, it is not truly 'multi-cellular'. The cells are really individuals, capable of existing on their own. In a truly multicellular organism the cells are specialized for different functions and are dependent on one another.

Each *Spirogyra* cell (Fig. 15.4) shows features typical of green plants:
1. a cellulose cell wall
2. one or more spirally coiled chloroplasts
3. pyrenoids, regions of starch formation and storage
4. a large vacuole.

Mucus on the outside of the cell prevents organisms settling and growing on the plant. The mucus also prevents cells drying out when trapped at the surface of the pond.

Nutrition: autotrophic, by photosynthesis.

Respiration: aerobic.

Gas exchange: simple diffusion across the cell wall. In sunlight, carbon dioxide diffuses in and oxygen diffuses out. At night, oxygen diffuses in and carbon dioxide diffuses out.

Locomotion: none. Drifts, carried by wind or water currents.

Sensitivity: none apparent.

Reproduction: asexual and sexual.

Asexual by binary fission (splitting in two). This occurs in spring and early summer when conditions are favourable. A long strand (filament) of identical cells results. Fragmentation occurs if the strand breaks.

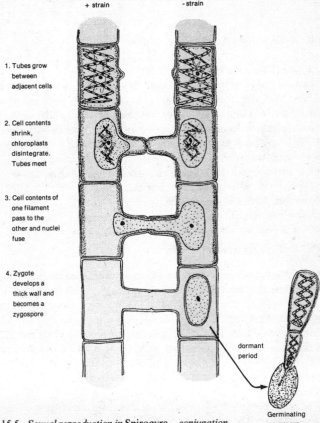

Fig. 15.5. *Sexual reproduction in* Spirogyra – *conjugation*

Sexual, by conjugation, between adjacent filaments (see Fig. 15.5). This occurs during periods of overcrowding and unfavourable conditions, such as occur in autumn. The end result is the production of a very resistant **zygospore** which is capable of overwintering. In conjugation, adjacent cells on two filaments form a conjugation tube between themselves. This

acts as a bridge through which the contents of one cell enter the other
In effect, the contents of each cell become 'gametes' which, when they
fuse, form a zygote. A hard capsule is then secreted, forming a zygospore
The zygospore falls to the bottom of the pond where it overwinters. The
zygospore may be dispersed to other ponds in mud on the feet of animals
In spring, the zygospore splits, releasing a *Spirogyra* filament of one or
two cells which by binary fission grows into a long filament.

15.8. *A Saprophytic Fungus:* Mucor

Phylum (Division): Fungi.

Habitat: the habitat *must* be moist. *Mucor* lives on the food it feeds
on, e.g. bread, jam, football boots.

Structure: *Mucor* consists of microscopic branching filaments called
hyphae (singular: hypha) which together form a network called a **mycelium**
(see Fig. 15.6). Each hypha is a tube of cytoplasm with a vacuole running
down the centre and surrounded by a cell wall composed of **chitin**. The
hyphae are unusual in that they are not divided up into cells (non-septate);
the hyphae are **multi-nucleate** (contain many nuclei).

Nutrition. *Mucor* is **saprophytic**, feeding on dead organic material by
secreting digestive enzymes on to it and then absorbing the soluble products of digestion. These products must be in solution, hence *Mucor*'s
dependence on a damp environment.

Respiration: aerobic.

Gas exchange: oxygen enters and carbon dioxide leaves by simple diffusion across the cell wall.

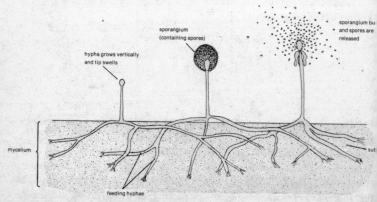

Fig. 15.6. *Asexual reproduction in* Mucor

Reproduction: both asexual and sexual reproduction occur.

Asexual reproduction (see Fig. 15.6): vertical hyphae develop **sporangia** (fruiting bodies) which each produce hundreds of tiny spores. In certain species of *Mucor* these spores are very resistant to drying out (desiccation) and can survive adverse conditions. The spores are very light and can be carried long distances on the wind. They will germinate under suitable damp conditions.

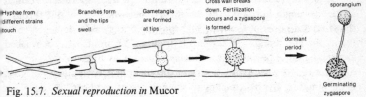

Fig. 15.7. *Sexual reproduction in* Mucor

Sexual reproduction (see Fig. 15.7): if hyphae of two different strains touch, **gametangia** (structures which produce gametes) may be formed at the point of contact. Nuclei (gametes) of the two strains fuse, to form a multi-nucleate zygote. As in *Spirogyra*, the zygote secretes a thick wall to form a **zygospore**. The zygospore remains dormant until favourable conditions exist, when it germinates to form a vertical hypha and sporangium which releases hundreds of spores.

15.9. *A Parasitic Fungus:* Phytophthora

Phylum (Division): Fungi.

Habitat: the fungus is a mould parasitic on the potato plant – it is known as the potato blight.

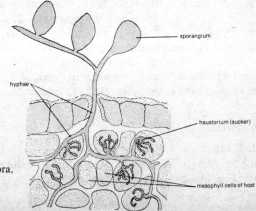

Fig. 15.8. Phytophthora, *in the leaf of a host*

Structure: like *Mucor, Phytophthora* consists of non-septate multi-nucleate hyphae. In addition, it has specialized suckers (**haustoria**) which penetrate its host's cells and by which it absorbs food.

Nutrition: *Phytophthora* obtains its food directly from the cells of the host plant. Sucker-like haustoria secrete digestive enzymes and enter and break down mesophyll cells.

Reproduction: both asexual and sexual (rare).

Asexual reproduction is similar to that of *Mucor*. Spores are formed inside **sporangia** borne on the end of hyphae which grow out through the stomatal pores of the leaf. The sporangium is easily detached from the hypha and remains intact, being dispersed by air currents or washed away by rain. The spores are not as resistant as those of *Mucor* and must germinate within a few hours, on another plant, if they are to survive. A germinating hypha attaches to the surface of the host and develops a fine tube which bores through the cuticle into an epidermal cell. The tube swells and branching hyphae grow out between the host cells and produce feeding haustoria.

Sexual reproduction: two hyphae from different strains grow together, one forming distinct male (antheridium) and the other female (oogonium) gamete-producing structures. One male gamete will fertilize a single female gamete to produce a zygote which develops into an **oospore**, a resistant overwintering form. Under suitable conditions, the oospore germinates, either producing a hypha which grows into host tissues, or forming sporangia as in the asexual method.

Control of potato blight

Uncontrolled spread of the potato blight fungus can devastate potato crops. The tubers (the edible part of the potato plant, see page 233) become infected when sporangia are washed down from the leaves. After being harvested, these tubers rot in storage. The Irish famine in the nineteenth century, in which millions of people died, was mainly caused by a widescale outbreak of potato blight.

Nowadays, various **fungicides** (chemicals which destroy fungi) are available. These usually contain heavy metals, e.g. copper, which are more poisonous to the fungus than to the plant.

Importance of fungi

1. **Useful:**
(a) Saprophytic fungi play a major role in recycling dead organic matter (see Figs 14.1 and 14.2, pages 289 and 290).

(b) Certain fungi produce antibiotics, e.g. *Penicillium* produces penicillin.

(c) Mushrooms and toadstools are fungi (albeit much larger than the moulds described here) and some are edible.

(d) Anaerobic respiration of yeasts is used to provide:

(i) alcohol, during fermentation to produce wine and other alcoholic beverages

(ii) carbon dioxide, to make the dough rise in bread.

2. Harmful:

(a) Decay fungi spoil food, e.g. *Penicillium* on jam, bread and cakes.

(b) Plant diseases such as potato blight are caused by parasitic fungi.

(c) Dry rot fungus destroys house timbers.

15.10. *The Male Fern:* Dryopteris

Phylum: Pteridophyta (the ferns)

Features of the group:

1. Leaves, roots and a stem are present.

2. A well-developed vascular system is present (for transporting substances up and down the plant).

3. The life cycle is complex and shows **alternation of generations**. An

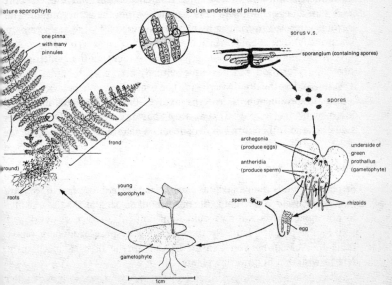

Fig. 15.9. *Life cycle of the fern*, Dryopteris

asexual sporophyte (spore-forming generation) alternates with a gameto-phyte (gamete-forming) generation. The sporophyte generation is dominant.

Habitat: widespread in woods, hedgerows and open heathland.

Structure: see Fig. 15.9. The **sporophyte** is the form by which most people recognize the fern. The leaves (fronds) have a feather-like appearance and bear many **pinnae** which are themselves divided into **pinnules**. On the underside of the pinnules are **sori** (singular: sorus). Each sorus contains many spore-forming **sporangia**. The underground stem (the rhizome) and roots anchor the fern to the ground. The rhizome stores food and the roots absorb water and minerals. In winter the fronds die back but the rhizome with its food store remains protected.

Nutrition: the fern is autotrophic. The fronds are green and contain chloroplasts for photosynthesis.

Reproduction and life history: in the sporophyte, mature sporangia dry out and explosively release their spores into the air. The spores are produced by meiosis (page 272) and are haploid. If a spore lands on suitable damp ground, it germinates to produce a small, green leaf-like structure, the **prothallus**. This is the gametophyte generation. It produces male organs (**antheridia**) which produce motile male gametes (sperms), and female organs (**archegonia**) which produce female gametes (eggs). Sperms are released and swim to and enter an archegonium, where one sperm fuses with an egg. Fertilization is external, within the archegonium. The sperm is attracted to the egg by chemicals secreted by the archegonium (chemotaxis). The zygote formed then grows into a new fern plant (sporophyte). To begin with, the young fern is dependent on the gametophyte (prothallus) for food, but the fern eventually becomes self-sufficient and is established as the dominant form. The prothallus withers away.

Sexual reproduction in the ferns involves *external* fertilization. In the flowering plants (see page 319) we see an advance over this – fertilization is *internal* and is therefore less dependent on chance.

Importance of alternation of generations

Terrestrial (land) plants such as mosses and ferns have developed this form of life cycle to overcome the problem of dispersion. As these plants do not move overcrowding would soon occur if all zygotes developed next to the parent plant. By incorporating a spore-producing phase in the life cycle, dispersion is ensured. Compare the dispersal mechanism in ferns with that in flowering plants.

15.11. *Flowering Plants (Angiosperms)*

Flowering plants can be grouped in a number of ways:

1. According to the number of seed leaves (cotyledons) they have:

(a) **Monocotyledons** (monocots for short) have one seed leaf. In the mature plant the vascular bundles are scattered throughout the stem and the leaves are narrow and parallel veined (see Fig. 15.10).

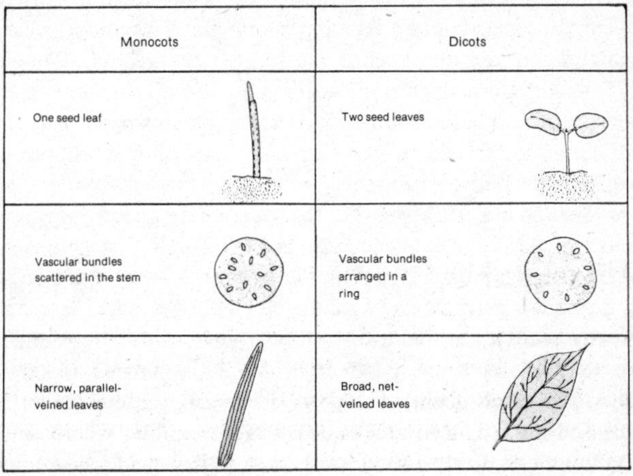

Monocots		Dicots	
One seed leaf		Two seed leaves	
Vascular bundles scattered in the stem		Vascular bundles arranged in a ring	
Narrow, parallel-veined leaves		Broad, net-veined leaves	

Fig. 15.10. *Monocotyledons and dicotyledons compared*

(b) **Dicotyledons** (dicots) have two seed leaves. This is the more common. In the mature plant the vascular bundles are arranged in a ring in the stem and the leaves are broad and net veined (see Fig. 15.10).

2. According to the duration of the life cycle:

(a) **Ephemerals** complete more than one life cycle in a year, overwintering as a seed. They are able to colonize new environments very rapidly. Many weeds are ephemerals, e.g. groundsel.

(b) **Annuals** complete one life cycle in a year, overwintering as a seed, e.g. broad bean.

(c) **Biennials** complete one life cycle every two years, overwintering as an underground food storage organ and then as a seed, e.g. carrot.

(d) **Perennials** complete one life cycle after several years. Woody perennials (see below) overwinter with trunk and branches remaining above ground. Herbaceous perennials (see below) overwinter as an underground food storage organ with (e.g. grass) or without (e.g. tulip) foliage above ground.

3. According to the degree of thickening in the stem:

(a) **Herbaceous** plants have unthickened soft green stems (see Fig. 15.11), e.g. buttercup, daisy.

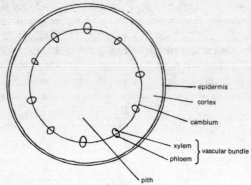

Fig. 15.11. *T.S. of a herbaceous dicot stem*

(b) **Woody** plants, shrubs and trees, have hard, brown, woody stems and roots. The 'wood' is xylem produced by **secondary** thickening; it provides the stem with support to bear the weight of branches and leaves. A continuous ring of meristematic (dividing) cells, the cambium, produces extra xylem on its inside and phloem on its outside (see Fig. 15.12). You will notice the xylem and phloem are not restricted to vascular bundles but are in a continuous ring or rings. Xylem forms growth rings, each one corresponding to a year's growth. A cork cambium below the epidermis gives rise to a layer of cork which is impermeable to water.

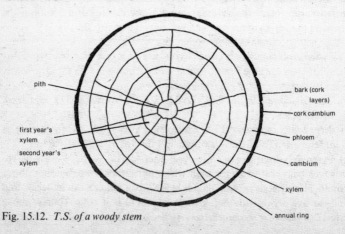

Fig. 15.12. *T.S. of a woody stem*

Woody plants which are **deciduous** lose their leaves in winter, e.g. oak, horse chestnut, whereas **evergreen** plants shed their leaves at irregular intervals, e.g. holly.

Structure of flowering plants

The parts of a typical herbaceous flowering plant are shown in Fig. 15.13. Flowering plants are made up of *three* main organs – stem, roots and leaves – and at certain times a fourth organ, the flower (page 240).

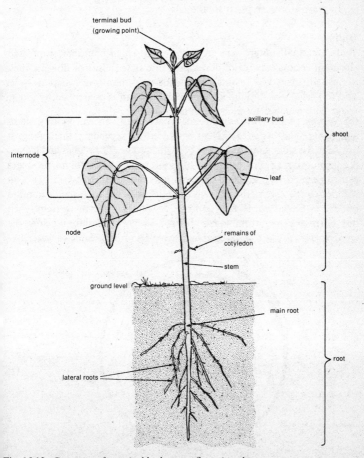

Fig. 15.13. *Structure of a typical herbaceous flowering plant*

The stem

The above-ground part of the plant is the shoot. The stem is the part of the shoot which bears the leaves and buds; it develops from the seed plumule (page 244).

The main functions of the stem are:

1. To hold the leaves aloft for photosynthesis (see page 64).
2. To hold the flowers aloft for pollination and seed dispersal (see pages 241 and 245).
3. To transport substances between the roots and leaves (see page 152).
4. In some cases, to store food (see page 234).
5. In green stems, to photosynthesize.

Buds

Buds are small outgrowths from the stem which contain compressed, dormant shoots capable of developing into new shoots or flowers; these are usually protected by bud scales. Buds are generally dormant in winter and in spring are stimulated to growth. Growth is controlled by auxins (page 209) which determine which buds develop when.

Twigs

Deciduous trees and shrubs shed their leaves at the onset of winter, leaving twigs of characteristic appearance (see Fig. 15.14).

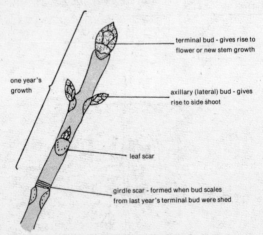

Fig. 15.14. *Horse chestnut twig in winter*

Roots

The main functions of roots are:

1. To anchor the plant in the ground.
2. To absorb water and dissolved mineral salts from the soil (see page 152).
3. To transport water and mineral salts to the stem (see page 152).
4. In some cases, to store food, e.g. the root tubers of the dahlia.

The structure of a young root tip is shown in Fig. 15.15.

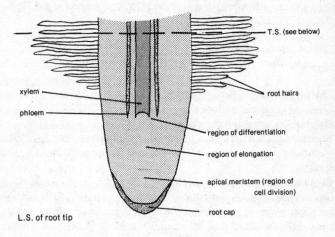

L.S. of root tip

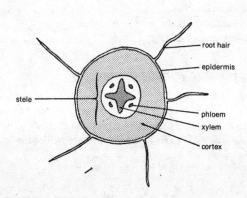

T.S. of root hair region

Fig. 15.15.

Leaves

These are attached to the stem at **nodes**.

The main functions of leaves are:

1. To carry out photosynthesis (see page 64).
2. To carry out gas exchange for respiration and photosynthesis (see page 129).
3. Transpiration (see page 154).
4. Amino acid synthesis (see page 68).
5. In some cases, to store food, e.g. the bulb of an onion (page 233).

15.12. *A Herbaceous Plant: the Bulbous Buttercup,* Ranunculus bulbosus

Phylum (division): Spermatophyta.

Class: Angiospermae (flowering plants).

Habitat: dry fields and meadows.

Structure: the buttercup is a herbaceous dicotyledon (see page 319).

Nutrition: autotrophic, by photosynthesis (see page 64) in the leaves and stem.

Respiration: aerobic (see page 107).

Gas exchange: by diffusion through stomata (see page 129). In sunlight, when photosynthesis exceeds respiration, carbon dioxide diffuses in and oxygen diffuses out. At night, respiration alone occurs, and oxygen diffuses in and carbon dioxide diffuses out.

Transport: see page 149.

Excretion: see page 165.

Sensitivity: see page 205.

Reproduction: sexual (see page 239).

The flower, pollination and fertilization

The buttercup flower (see Fig. 11.2, page 240) is hermaphrodite and insect-pollinated. The flower is bright yellow, scented, has nectar, and attracts a variety of insects such as flies, bees and beetles. To help prevent self-pollination, the male parts (stamens) mature before the female parts (stigmas).

Fruit formation and seed dispersal

After pollination and fertilization, each fertilized ovule forms a seed surrounded by a leathery fruit case. The seeds stay attached to the receptacle which breaks off to form a light bunch of fruits which is dispersed by the wind.

Germination

Germination is **hypogeal**, the cotyledons remaining below ground (see page 249).

This particular species of buttercup is **perennial** (see page 319). It overwinters as an underground food storage organ.

15.13. *A Deciduous Tree: the Horse Chestnut,* Aesculus hippocastanum

Phylum (division): Spermatophyta.
Class: Angiospermae (flowering plants).
Habitat: municipal parks and edge of fields.
Structure: see Fig. 15.16. The horse chestnut is a woody dicotyledon (see page 320).
Nutrition: autotrophic, by photosynthesis (see page 64) in the leaves. The stem is covered in bark and does not photosynthesize.
Respiration: aerobic (see page 107).
Gas exchange: by diffusion through stomata and lenticels (page 129).
Reproduction: sexual (see page 239).

The flower, pollination and fertilization

The flowers open in May. The 'candle-like' inflorescence (flower arrangement) consists of many single flowers attached to an upright shoot (see Fig. 15.16). The flowers are insect-pollinated. Bees are attracted by the white petals with orange bases and nectaries. Both male only and hermaphrodite flowers may be found in a single inflorescence. Within a hermaphrodite flower, the female parts (stigmas) mature before the male parts (stamens) to minimize self-pollination.

Fruit formation and seed dispersal

After pollination and fertilization, each fertilized ovule forms a very large
seed (a conker) surrounded by a spiny, leathery fruit case (see Fig. 15.16).

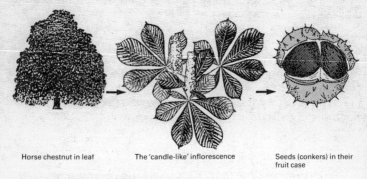

Horse chestnut in leaf   The 'candle-like' inflorescence   Seeds (conkers) in their
                                                                        fruit case

Fig. 15.16.   *The horse chestnut*

The seed may germinate where it falls or may be dispersed by animals
either carrying or eating the seed. Large mammals such as deer and cattle
may carry fruit cases trapped in their fur or may eat the seed which
occasionally passes through the digestive system unharmed. Squirrels
burying conkers is a common way in which seeding takes place.

Germination

Germination is **hypogeal**, the cotyledons remaining below ground (see
page 249).

15.14. *A Protozoan:* Amoeba

Phylum: Protozoa.
Habitat: still or slow-moving fresh water.
Structure: see Fig. 15.17. Up to 1 mm in diameter. Irregular shape,
constantly changing. A single-celled (unicellular) animal.
Animal features:
1. no cell wall.
2. contractile vacuole (for osmoregulation).
3. food vacuoles (see nutrition below).
Nutrition: holozoic. Feeds on bacteria (page 307), algae such as *Chlamydomonas* (page 310) and other protozoa. Ingests its prey by **phagocytosis**

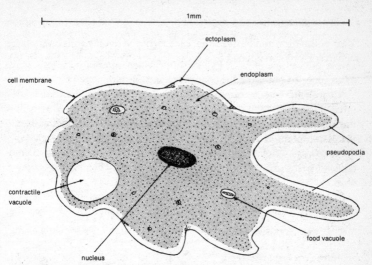

Fig. 15.17. Amoeba proteus

(Fig. 15.18) using pseudopodia. Digestive enzymes are secreted into the food vacuole, the soluble products of digestion absorbed, and indigestible food egested.

Respiration: aerobic.

Fig. 15.18. *Feeding in* Amoeba

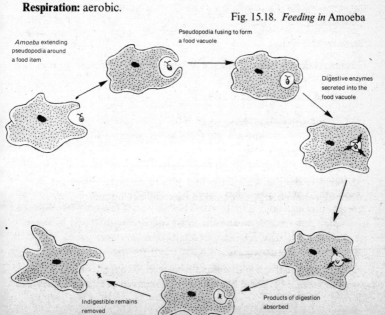

Gas exchange: oxygen enters and carbon dioxide leaves by simple diffusion across the cell membrane (see page 119).

Excretion and osmoregulation: soluble nitrogenous wastes diffuse out across the cell membrane. Excess water which enters by osmosis (page 57) is pumped out using the contractile vacuole (see page 167).

Movement: *Amoeba* moves by pushing out pseudopodia ('false feet') into which it flows. Liquid cytoplasm (plasmasol) is thought to flow through a funnel of jelly-like cytoplasm (plasmagel) to form and extend the pseudopodium.

Sensitivity: follows the trail of chemicals released by its prey, (chemotaxis). Moves away from strong light, harmful chemicals and sharp objects.

Reproduction: asexual only. On reaching a limiting size, it divides in two (binary fission) to produce two identical daughter cells (see Fig. 15.19).

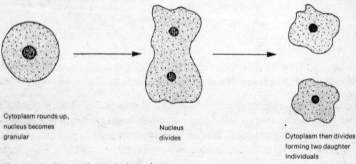

Cytoplasm rounds up, nucleus becomes granular

Nucleus divides

Cytoplasm then divides forming two daughter individuals

Fig. 15.19. *Binary fission in* Amoeba proteus

15.15. *A Coelenterate:* Hydra

Phylum: Coelenterata.

Habitat: still or slow-moving fresh water, usually attached to vegetation.

Structure: see Fig. 15.20. *Hydra* is about 1 cm long when fully extended. It has a simple sac-like body with one opening which serves as both mouth and anus. This leads into a digestive cavity, the enteron. Outside the mouth is a ring of tentacles. The body wall has only two layers of cells, the outer **ectoderm** and the inner **endoderm**, separated by the jelly-like **mesogloea**. The ectoderm contains muscle, sensory and sting cells. The cells of the endoderm are specialized for digestive functions. The mesogloea contains a network of interconnected nerve cells (a nerve net).

Hydra is at the **tissue level** of organization; life processes are mainly

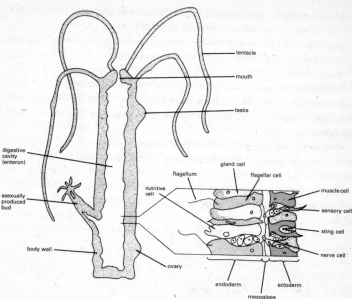

Fig. 15.20. *V.S. of* Hydra *with detail showing body wall*

carried out by tissues rather than organs. It is arguable whether tentacles are organs; the ovaries and testes are sometimes regarded as organs.

Nutrition: holozoic. *Hydra* preys on small aquatic organisms such as water fleas. When a flea comes in contact with a tentacle, sting cells discharge their barbs which penetrate and paralyse the flea. The prey is **ingested** when the tentacles bend over and drop the flea into the enteron. Digestion is both **extra-** and **intracellular**. Gland cells secrete digestive enzymes into the enteron. Nutritive cells ingest partially digested food by phagocytosis; further digestion occurs inside the cells. Soluble products of digestion are absorbed directly by all cells in the body. There is no transport system, although flagella do help circulate the contents of the enteron. Undigested material is egested via the mouth.

Some *Hydra* have green algae living symbiotically within their endoderm cells. *Hydra* gains food and oxygen produced by the photosynthesizing algae, while the algae gain carbon dioxide and nitrogenous wastes from *Hydra*.

Respiration: aerobic.

Gas exchange: by simple diffusion across the body surface.

Movement: *Hydra* is generally sessile (stays in one place) but it can move by 'somersaulting' on its tentacles and then reattaching. Muscle strands run both longitudinally and in a circular direction; contraction of the former shortens and fattens the body while contraction of the latter lengthens and narrows the body. Coordination of movements is achieved using the nerve net which connects sensory cells with muscle cells.

Reproduction: both asexual and sexual reproduction occurs.

Asexual reproduction occurs under favourable conditions when *Hydra* develops outgrowths, or buds, from the body wall. A bud grows to form a daughter *Hydra* which eventually separates from the parent.

Sexual reproduction occurs when living conditions are unfavourable. Most *Hydra* species are hermaphrodite, but testes and an ovary are formed at different times, so preventing self-fertilization. Within the ovary, the egg is fertilized by the sperm from another *Hydra*. The zygote formed secretes a resistant case around itself and drops to the bottom of the pond. It remains dormant until favourable conditions occur, when the case splits, releasing a hollow ball of cells which grows into a new *Hydra*.

15.16. *A Parasitic Flatworm:*
the Pork Tapeworm, Taenia solium

Phylum: Platyhelminthes (the flatworms).

Habitat: the pork tapeworm lives in the intestine of man. It is an obligate parasite (it dies if deprived of its host).

Structure: see Fig. 15.21. The body is a creamy white ribbon up to 5 metres long, 5 mm wide and 1 mm thick. At the anterior end, the scolex is attached by hooks and suckers to the lining of the host's small intestine. Just behind the scolex hundreds of proglottides are formed which move back along the ribbon as they develop.

Nutrition: no digestive system. Products of human digestion are absorbed directly across the body wall.

Respiration: anaerobic (little oxygen is present in the intestine and as the worm is relatively inactive it does not have a high energy demand, see page 107).

Movement: some muscle tissue is present but the tapeworm shows only slight wriggling movements.

Sensitivity: no sense organs apparent, but a simple nervous system is present.

Reproduction and life cycle: see Fig. 15.21. In common with many other parasites, a large amount of energy is invested in reproduction. The tapeworm is hermaphrodite and self-fertilization occurs (the probability of

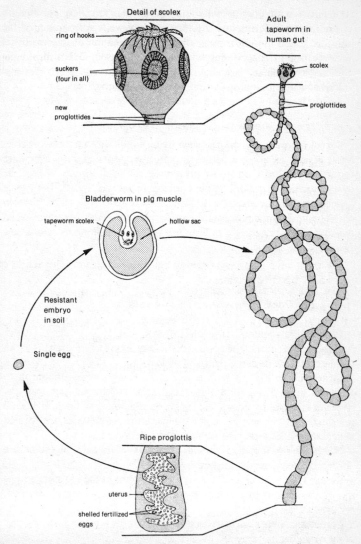

Fig. 15.21. *Life cycle of pork tapeworm*, Taenia solium

encountering another tapeworm is very low). As each proglottis matures it produces first sperms and then eggs. Ripe proglottides each containing thousands of fertilized eggs detach from the posterior end of the worm and pass out of the host's body in the faeces. The eggs contained in the released proglottides will develop only if eaten by the **secondary host**, the pig. Inside the pig a small worm, the bladderworm, develops and embeds itself in skeletal muscle. Should the infected pork then be eaten by man, the **primary host**, then the bladderworm may reach the small intestine where it develops into the adult tapeworm.

Features associated with its parasitic way of life

1. A scolex for secure attachment to host.
2. A flattened body to increase the surface area for absorption of digested food.
3. Simple body structure (loss of unnecessary organs):
(a) no digestive system
(b) poorly developed muscle
(c) no sensory organs.
4. Anaerobic respiration.
5. A thick cuticle covers the body and protects against the action of the host's digestive enzymes.
6. The worm is hermaphrodite and self-fertilization occurs (see reproduction above).
7. The chance of an egg reaching a new host is very small and so large numbers of eggs are produced continually. A worm may produce over 1,000 million eggs in a lifetime.
8. The eggs are highly resistant and can withstand long periods in the soil.
9. The use of a secondary host increases the chance of eventual entry into the primary host (pork is eaten by man).

Methods of control

The pork tapeworm is very rare in Britain and is common only in those countries where sanitation and standards of hygiene are poor. The following methods are used to control the parasite.
1. Good sanitation prevents untreated sewage from reaching the soil where pigs can ingest eggs.
2. Proper inspection of meat for human consumption – pork containing bladderworms has a spotted 'measly' appearance.
3. Thorough cooking of pork kills the bladderworms.

15.17. *A True Worm: the Earthworm,* Lumbricus terrestris

Phylum: Annelida (true worms).

Habitat: burrows in damp soil up to 2 metres down, coming to the surface at night or during heavy rain.

Structure: see Fig. 15.22. Up to 30 cm long, the body is divided into distinct segments. Essentially the body consists of two tubes, one inside the other; the body wall on the outside, the alimentary canal on the inside, and the two separated by a fluid-filled cavity which acts as a hydrostatic skeleton (see page 215). There is a true gut with mouth and anus. Bristles called **chaetae** can be extended from the ventral surface and are used in locomotion.

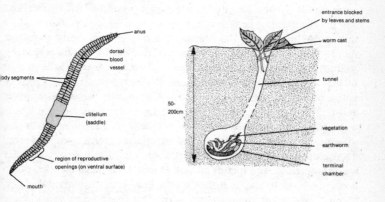

Fig. 15.22. (*a*) *Dorsal view of earthworm* (*b*) *Earthworm in its burrow*

Nutrition: holozoic. Vegetation and soil particles are ingested through the mouth and pass along the gut under muscle action. Digestive enzymes are secreted by the gut lining and the soluble products of digestion are absorbed into blood capillaries for transport to all cells in the body. Undigested food and soil particles pass out via the anus to form the familiar worm cast.

Respiration: aerobic.

Gas exchange: like *Amoeba* and *Hydra*, gas exchange in the earthworm takes place by simple diffusion across the body surface. In the earthworm, however, a circulatory system is necessary to deliver the oxygen to all internal tissues. The skin of the earthworm is a specialized respiratory surface (see page 119).

Movement: changes in body shape are brought about by the action of antagonistic circular and longitudinal muscles acting on the fluid skeleton. When circular muscles contract, the body becomes thin and long; when longitudinal muscles contract, the body becomes short and fat. During forward movement through the burrow or on the soil surface, alternate contractions and relaxations start at the head and pass back along the body. The head is extended, then fattens, with its chaetae anchored in the soil, and then the rest of the body is drawn up towards it. If the soil is tightly compacted, soil is steadily ingested as the worm moves forwards. The worm produces mucus (slime) to lubricate the sides of the burrow.

Reproduction: the earthworm is hermaphrodite, but cross-fertilization occurs (two individuals exchange sperm). Mating takes place at the soil surface, usually at night, and is a complex process which takes about 3 hours. Two worms lie with their ventral surfaces together, but pointing in opposite directions. The clitellum (saddle) of each worm secretes a mucus tube which binds the two together. Sperms are exchanged by passing from the male opening of one worm into grooves on the surface of the other worm. The grooves lead to sperm storage sacs. Tubes of mucus prevent the sperms from mixing. After mating, the worms separate and return to their burrows. The clitellum secretes a substance which will later harden to form a cocoon. A tube of this substance is drawn towards the anterior (head) end and as it passes over the female repro-ductive opening several eggs are deposited in it. Next it passes over the sperm storage sac and sperms are passed into it. Fertilization takes place inside the cocoon, and as it slips off the head end, the ends seal. After about 12 weeks, *one* young worm emerges from the cocoon.

Importance of earthworms

By their actions earthworms improve the fertility of soil:
1. Their burrows aerate, drain and break up the soil.
2. Their casts, which contain fine soil particles, help produce a good tilth (seed bed). Their action also mixes soil from lower layers with that on the surface, thus increasing the depth of topsoil.
3. The worm's waste products add manure to the soil and encourage the activity of soil bacteria.
4. Pieces of vegetation, e.g. leaves, dragged down into burrows help to mix humus with the soil.

15.18. *An Insect with Complete Metamorphosis: the Cabbage White Butterfly,* Pieris brassicae

Phylum: Arthropoda (arthropods).
Class: Insecta (the insects).
Insects have the following features:

1. The body is divided into three parts: the head, thorax and abdomen.
2. Three pairs of jointed legs are attached to the thorax.
3. Usually two pairs of wings are attached to the thorax.
4. The body is covered in a cuticle (exoskeleton) made of chitin.
5. A pair of antennae and compound eyes are present on the head.
6. Insects exchange gases through a system of air tubes called **tracheae**.

Habitat: migratory, but can be seen in gardens and rural regions during warm, dry weather.

Structure: see Fig. 15.23 (overleaf).

Nutrition: holozoic. In the adult, the nectar of flowers is the main food and is sucked up using an extensible coiled **proboscis**, a tube-like structure formed from mouth parts called **maxillae**. The caterpillar (the larval phase) feeds on cabbage leaves and has biting mouthparts, the **mandibles**.

Respiration: aerobic.

Gas exchange: by diffusion through a system of tracheae and tracheoles (see page 120).

Reproduction and life cycle: see Fig. 15.23. The cabbage white butterfly shows **complete metamorphosis** (see page 267). The first adults emerge from pupae in late April. They live for about three weeks during which time they mate; fertilization is internal. The female lays up to 100 eggs on the underside of cabbage leaves. After about a week the eggs hatch and small caterpillars emerge. They feed actively on the cabbage leaves and grow very rapidly, reaching 5 cm long in a month, during which time they moult four or five times. The full-grown caterpillar climbs a post, wall or tree and pupates a metre or two above ground. During pupation, the larval organs break down and are replaced by adult organs. After 3 or 4 weeks the pupal case splits open and the adult emerges. Adults then mate, the females lay eggs and the second generation of pupae generally overwinter with the adults emerging the following spring.

Importance to man: the cabbage white butterfly can cause local damage to vegetable crops. Its numbers are controlled by a parasitic wasp which lays eggs inside the caterpillars. When the eggs hatch the wasp larvae feed on and destroy the caterpillar.

For certain boards you are required to know details of the life history of an insect with **incomplete metamorphosis** e.g. the locust or cockroach

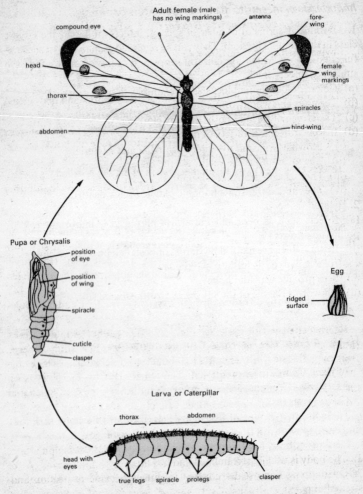

Fig. 15.23. *The life cycle of the cabbage white butterfly*, Pieris brassicae

(see page 267); on other boards you are required to know the life cycle of a social insect, e.g. the bee. Check with your syllabus.

Importance of insects to man

1. **Harmful:** in terms of number of species, insects are by far the largest class in the animal kingdom. They can be regarded as man's most successful competitors. Many agricultural crops and manufactured goods are subject to insect attack:

(a) Colorado beetles and locusts, for example, destroy crops.

(b) Termites and woodworm attack wood.

(c) The larvae of certain moths destroy clothing.

(d) Insects transmit a variety of diseases:

   (i) Blowflies transmit dysentery protozoa and bacteria.

   (ii) The mosquito is the vector for the malaria parasite.

   (iii) The tsetse fly transmits sleeping sickness.

   (iv) The rat flea transmits bubonic plague (black death).

2. **Helpful:**

(a) Insects such as bees and butterflies play a very important role in pollinating flowers.

(b) Insects which are natural predators of pests can be used for biological control, e.g. ladybirds feeding on aphids.

(c) Bees provide wax and honey.

(d) Silkworms (larvae of the silkworm moth) provide silk.

15.19. *A Fish: the Herring,* Clupea harengus

Phylum: Chordata (chordates).
Sub-phylum: Vertebrata (vertebrates).
Class: Pisces (fish).

There are two groups of fish: bony fish (those with a bony skeleton) and cartilaginous fish (those with a skeleton of cartilage).

Features of bony fish (most are adaptations for an aquatic life):

1. The body is usually streamlined and has fins.

2. Overlapping bony scales covered in mucus provide protection and streamlining.

3. The organs of gas exchange are gills.

4. In most, movement is brought about by waves of contraction that pass along muscle blocks (myotomes) on either side of the bony vertebral column (see page 225).

5. In most, a lateral line is used for detecting vibrations on the water.

6. Fish are poikilothermic (have a variable body temperature (see page 175)).

7. Sexes are separate and fertilization is usually external.

Habitat: marine, lives in waters around the British Isles.

Structure: see Fig. 15.24. Examine the external features as adaptations for aquatic life.

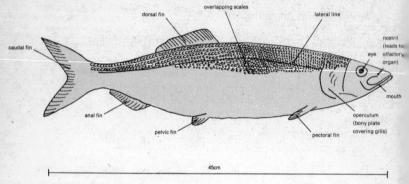

Fig. 15.24. *The herring,* Clupea harengus

Nutrition: holozoic. They are carnivorous, feeding mainly on copepods (small crustacean arthropods) which they filter out of the water using gill rakers.

Excretion and osmoregulation: water leaves the fish by osmosis (page 57) and is replaced by drinking seawater and then actively expelling the salts it contains (see page 167). Soluble nitrogenous wastes are excreted in the urine produced by the kidneys.

Respiration: aerobic.

Gas exchange: the site of gas exchange is the gill eipthelium which is ventilated (see page 121).

Locomotion: see page 225. The fins are used for steering, propulsion and provide stability.

Reproduction and life history: spawning occurs in autumn on stony sea beds in shallow coastal waters. Fertilization is external; females shed ribbons of eggs on to the sea bed and males then release sperm over the eggs. A single female can lay up to 20,000 eggs per year but few will develop to maturity. The chance of fertilization and survival of the eggs is low but is counteracted by the sheer numbers which are laid (see page 238). The larvae hatch out after about 2 weeks and depend initially on a yolk sac for their food supply. Soon after, they swim up into the surface waters where they feed and remain for 3–6 months while they develop into adults.

NOTE. On certain syllabuses you are required to know the reproductive behaviour of the stickleback, *Gasterosteus aculeatus*.

15.20. *An Amphibian: the Common Frog,* Rana temporaria

Phylum: Chordata (chordates).
Sub-phylum: Vertebrata (vertebrates).
Class: Amphibia (amphibians).
Features of the group:
1. Poikilothermic vertebrates (see page 175) with paired limbs.
2. Adult terrestrial, larva aquatic.
3. Adult has lungs for gas exchange, larva has gills.
4. Skin of adult usually thin, moist and mucus-covered.
5. Adults return to water to breed and eggs are laid and fertilized in the water.
6. Gradual **metamorphosis** (see page 267) from larva to adult occurs.
   Habitat: moist, shady places or in ponds and streams.
   Structure: see Fig. 15.25.

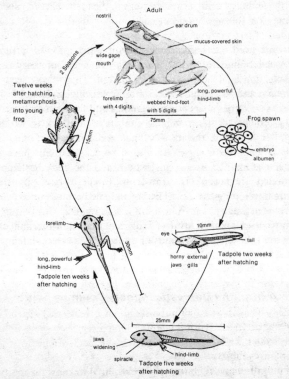

Fig. 15.25. *Life cycle of the common frog*

Nutrition: holozoic. Adult feeds mainly on slugs, snails and insects which are caught by the rapid ejection of the frog's long sticky tongue. In the young larva (tadpole) a yolk sac provides the food supply till the mouth develops. Once the mouth is formed, the tadpole feeds on the covering of algae growing on the surface of stones and large plants.

Respiration: aerobic.

Gas exchange: newly hatched tadpoles have external gills for gas exchange. 2 to 3 weeks after hatching the gills become internalized and water is pumped over them between mouth and spiracle (see Fig. 15.25). In the adult, the mouth, lungs and skin are used for gas exchange (see page 122).

Locomotion: the hind-limbs of the adult are powerful for jumping on land and for swimming, assisted by the webbed feet. The short forelimbs act as shock absorbers on landing and are steering aids when swimming.

Reproduction and life history: frogs mate in early spring, the male on top of the female. Fertilization is external, the male releasing sperm over what may be 100 newly laid eggs. Coupling in this way increases the likelihood of fertilization.

The eggs swell soon after contact with water and the fertilized egg becomes surrounded by 'jelly'. The 'jelly' provides buoyancy and helps give protection against injury and predation. The embryo feeds on the jelly and hatches out about 2 weeks after fertilization. The newly hatched tadpole has external gills for gas exchange and feeds on an internal yolk supply. A sucker on its underside enables it to attach to weeds. By 2 weeks after hatching, the mouth, anus and gut have formed and the tadpole is feeding on algae. By about 5 weeks the gills have become enclosed, the hind-limbs appear as buds and the body has lengthened. By 10 weeks, the tadpole is carnivorous, forelimbs have appeared, and lungs are developing. At 12 weeks, the tadpole becomes immobile, sheds its larval skin and metamorphoses into a young frog. It takes two more seasons to reach sexual maturity. Frogs hibernate in winter, under stones or in mud at the bottom of ponds or ditches.

15.21. *A Bird: the Domestic Pigeon,* Columba livia

Phylum: Chordata (chordates).
Sub-phylum: Vertebrata (vertebrates).
Class: Aves (birds).
Features of the group (adaptations for flight):
1. Forelimbs developed to form wings.

2. Feathers provide a lightweight protective and insulating layer on the body.

3. No teeth (instead, a beak).

4. Homoiothermic (see page 175) – a constant high body temperature is necessary to enable high rate of respiration to provide energy during flight.

5. Highly developed nervous system enables remarkable muscular co-ordination.

Habitat: common in towns and cities.

Structure: see Fig. 15.26.

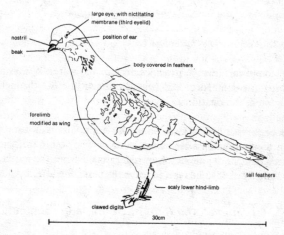

Fig. 15.26. *The domestic pigeon,* Columba livia

Nutrition: holozoic. Feeds on seeds, young green shoots and occasionally insects and snails.

Respiration: aerobic.

Gas exchange: ventilated lungs and air sacs are used. An efficient system is required because of the high rate of respiration during flight.

Locomotion: flapping flight, see page 227.

Reproduction: fertilization is internal, the male mounting the female after courtship display. Sperm from the male is deposited in the oviduct of the female. Yolk, albumen, membranes and a hard shell develop around the fertilized egg. The resulting hard-shelled 'egg' (Fig. 15.27) continues down the oviduct and is laid in a nest constructed by both parents. The developing embryo depends on yolk for its carbohydrate

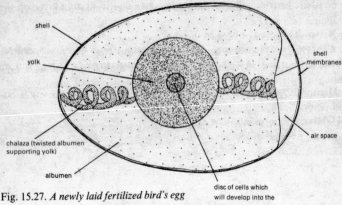

Fig. 15.27. *A newly laid fertilized bird's egg*

and fat supplies; albumen provides protein and water. Wastes are either deposited in a solid form, e.g. uric acid, or diffuse out through the egg shell, e.g. carbon dioxide. Oxygen is obtained by diffusion through the egg shell.

The egg is incubated (kept warm) by both parents in turn. The two eggs in a clutch hatch after about 18 days. The young feed initially on 'pigeon's milk', a white secretion regurgitated from the mother's crop. The young fly at 35–37 days and leave the nest soon after.

15.22. *A Mammal: the Rabbit,* Oryctolagus cuniculus

Phylum: Chordata (chordates).
Sub-phylum: Vertebrata (vertebrates).
Class: Mammalia (mammals).
Main features of the group:
1. Body surface covered in hair.
2. Homoiothermic (constant body temperature).
3. Give birth to live young (viviparous).
4. Feed their young on milk, from mammary glands.
   In addition they have the following features:
1. Large cerebrum of the brain (see page 186).
2. External ear, the pinna, is present (see page 195).
3. There are three ear ossicles in the middle ear (see page 196).
4. Heterodont dentition – different types of teeth are present (see page 98).

5. A hard palate separates the nasal passages from the buccal cavity or mouth (see page 83).

6. Sweat glands are present in the skin (see page 174).

7. A diaphragm separates the thorax from the abdomen (see page 123).

Habitat: burrows in light, sandy soil in open woods or grassland. Communal burrows consist of connecting tunnels with several entrances.

Structure: see Fig. 15.28.

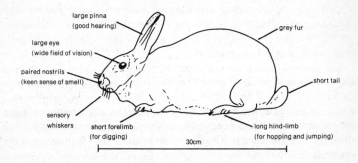

large pinna
(good hearing)

grey fur

large eye
(wide field of vision)

paired nostrils
(keen sense of smell)

short tail

sensory
whiskers

short forelimb
(for digging)

long hind-limb
(for hopping and jumping)

30cm

Fig. 15.28. *The rabbit,* Oryctolagus cuniculus

Nutrition: holozoic. Herbivorous (see page 81). Feeds on grass seeds, leaves, young bark.

Reproduction and life history: under favourable conditions, 3–6 litters each containing 3–7 young are born between March and September. Gestation takes about 4 weeks and the young are born blind. The young are suckled for about four weeks and mature at around 6–8 months. For most individuals, because of predation, the life span is around one year.

Their very high reproductive rate, resulting from short gestation, many litters and early sexual maturity sometimes makes rabbits a pest on agricultural land. The virus disease myxomatosis was introduced to control them, but some rabbits are now immune to this disease.

16. In the Exam

If you followed the advice and instructions in Chapter 2 during your revision, you should now be familiar with the different types of question set by your board and should know how to tackle them. In any case, read over Chapter 2 once more.

Get plenty of sleep the night before your exam. Do *not* stay up all night revising. Get up in good time for the exam, allowing extra time for travel in case of unforeseen delays (traffic jams, train cancellations, a bicycle puncture, etc.). Turn up to the exam in good time and come fully equipped with pen, pencils, pencil sharpener, eraser and ruler.

Now is the time to put all your effort to good use. Remember, what counts is what you put down on the exam paper. This is your only contact with the examiner. Marks can only be given for what you write on the paper – not for what remains in your head.

1. Read carefully the instructions at the head of the exam paper and follow them absolutely.

2. Work out a time allocation for the questions according to the marks they carry. Throughout the exam, check on the time and stick to your time allocation – do not overrun.

3. Answer all compulsory questions.

4. If the paper is divided into sections make sure you answer the correct number of questions from each section.

5. Always answer the correct *number* of questions. If you are required to do four long-answer questions and you only answer three, it is almost impossible to make up the marks for the missing fourth question. The first 50 per cent of marks are the easiest to gain on any question. To get 60 marks out of a 100, it is easier to gain 15 out of 25 on each of four questions than to gain 20 out of 25 on each of three.

6. If given a choice of longer-answer questions, narrow down the best questions to answer. You can do this by estimating the number of marks you will get for a question *before* you attempt to answer it.

7. *Understand* what the question asks *before* you attempt to answer it. Do *not* twist the question into an unintended meaning. *No* marks are awarded for irrelevant information and the answer should include *only*

points demanded by the question. Do not write down *all* you know on a particular topic unless it is asked for.

8. Examiners prefer concise answers written in good English. Set out work neatly – the examiner is only human. Draw a single straight line across any work which you do not wish to be marked (an essay plan for example).

9. Plan your longer answers before writing them (see page 24).

10. Use large labelled diagrams if they make your answer clearer and save time and words.

11. If you are running out of time in a question, complete your answer in note form, putting in all the key points. You will gain at least some marks.

12. If any time is left at the end, read through your answers and correct any mistakes.

Look on the exam as a culmination of your efforts, a pinnacle you have reached rather than a dreadful hurdle you must haul yourself across.

Answers to Multiple-choice Questions

Chapter 3: Q. 5 (page 54): D.
Chapter 5: Q. 5 (page 104): A; Q. 6 (page 104): E.
Chapter 7: (page 139): (a).
Chapter 8: Q. 2 (page 179): D.
Chapter 14: (page 290): D; Q. 1 (page 299); (a) D; (b) A; (c) D; (d) B; (e) B

MORE ABOUT PENGUINS, PELICANS AND PUFFINS

For further information about books available from Penguins please write to Dept EP, Penguin Books Ltd, Harmondsworth, Middlesex UB7 0DA.

In the U.S.A.: For a complete list of books available from Penguins in the United States write to Dept DG, Penguin Books, 299 Murray Hill Parkway, East Rutherford, New Jersey 07073.

In Canada: For a complete list of books available from Penguins in Canada write to Penguin Books Canada Ltd, 2801 John Street, Markham, Ontario L3R 1B4.

In Australia: For a complete list of books available from Penguins in Australia write to the Marketing Department, Penguin Books Australia Ltd, P.O. Box 257, Ringwood, Victoria 3134.

In New Zealand: For a complete list of books available from Penguins in New Zealand write to the Marketing Department, Penguin Books (N.Z.) Ltd, P.O. Box 4019, Auckland 10.

In India: For a complete list of books available from Penguins in India write to Penguin Overseas Ltd, 706 Eros Apartments, 56 Nehru Place, New Delhi 110019.

Penguin Examination Bestsellers

D. H. Lawrence/The Rainbow
D. H. Lawrence/Sons and Lovers
Laurie Lee/Cider With Rosie
Jack London/The Call of the Wild and Other Stories
Gavin Maxwell/Ring of Bright Water
George Orwell/Animal Farm
George Orwell/ Nineteen Eighty-Four
Alan Paton/Cry, the Beloved Country
Jonathan Swift/Gulliver's Travels
Flora Thompson/Lark Rise to Candleford
Mark Twain/Huckleberry Finn
Keith Waterhouse/Billy Liar
Evelyn Waugh/Brideshead Revisited
H. G. Wells/Selected Short Stories
John Wyndham/The Day of the Triffids